# Dynamic Water-System Control

*Design and Operation of*
*Regional Water-Resources Systems*

ARNOLD H. LOBBRECHT
*DHV Water BV*
*Delft University of Technology, Faculty of Civil Engineering*

A.A.BALKEMA/ROTTERDAM/BROOKFIELD/1997

Keywords: Water management; water resources; control system; real-time control; dynamic control; optimization; successive linear programming; interests; strategy; design.

Published by
A.A. Balkema, P.O. Box 1675, 3000 BR Rotterdam, Netherlands
Fax: +31.10.4135947; E-mail: balkema@balkema.nl; Internet site: http://www.balkema.nl

A.A. Balkema Publishers, Old Post Road, Brookfield, VT 05036-9704, USA
Fax: 802.276.3837; E-mail: info@ashgate.com

ISBN 90 5410 431 7

*To my parents*

# Table of Contents

# Preface

Dynamic control of water systems is an challenging new field of study, which is currently attracting a lot of attention. Water managers, engineers and advisors are trying to develop methods to achieve better management of water systems, and meet ever higher quantity, quality and safety requirements. The aim of this thesis is to make a contribution to these developments.

The study which is reported in this thesis should not be seen in isolation, but must be considered the next step in a series of developments which took place mainly in Europe during the past decade, especially in sewer-system control. A regional water system comprises various subsystems, one of which is the sewer system. Consequently, regional water systems are more complex and several water authorities may be involved in management. This complexity has led to the development of a methodology, which considers the entire water system, as well as all requirements of different interest groups that exist in that system.

The research took place at Delft University of Technology, in the Department of Water-Management, Environmental and Sanitary Engineering, Section Land and Water Management and was co-funded by a number of institutions: the Foundation for Applied Water Research (STOWA), the Water Boards Delfland and De Drie Ambachten, and DHV Water BV.

The present study focuses, in addition to theory development, on practical applications of the new methodology. The results will hopefully be incorporated in many water managers' daily practice.

The general awareness that we should aim for a sustainable living environment, where sufficient water of the right quality is continually available, is currently reflected in various developments in integrated water management. These developments will hopefully persuade others to attempt further fine-tuning of methods which will improve the quality of our water-related environment. Not in the least because it is a challenging field, which would benefit from improved efficiency: better management at a lower cost.

Arnold Lobbrecht

# Acknowledgments

In the first place I wish to thank my promotors, Wil Segeren of the Faculty of Civil Engineering and Freerk Lootsma of the Faculty of Technical Mathematics and Informatics. Both of them have greatly encouraged me to develop new ideas while their advice and constructive criticism have been invaluable. I am especially grateful for the loyalty of Wil Segeren, my main supervisor, and for the freedom he allowed me during the almost four years this research into dynamic control in water management has taken.

My employer, DHV Water BV is thanked for giving me the opportunity to spend the time required for this research. I am especially indebted to Willem Witvoet for his support throughout this study.

STOWA's substantial funding made this study possible, the practical assistance of Ludolph Wentholt deserves a special mention in this respect.

The research, presented in this thesis, benefitted greatly from the dedication of a team of researchers and MSc students at Delft University of Technology. As part of their final MSc assignments the following persons helped develop the methodology described and test its practical applications: Ton Botterhuis, Alex Hoogendoorn, Jean Philippe Janssens, Erik Schuilenburg, John Steenbekkers, Rudolf Versteeg and Tony Vredenberg. After graduating, Alex, John and Paul Willem Vehmeyer, stayed on as research associates to help develop a new Decision Support System: AQUARIUS.

Kick Bouma, Hans Hartong and Peter Paul Verbrugge reviewed parts of the drafts from a practical point of view. My friends Willem Mak and Tony Vredenberg proved to be of great help in giving detailed comments on the final draft. The English of the draft text was edited in minute detail by Thea van de Graaff. Hopefully, some of her work has survived the many revisions that resulted in this final version.

Finally, and above all, I want to thank my wife, Annette and sons, Caspar, Guido and Victor for their patience and understanding during these past few years for a family member who increasingly withdrew into himself and took no notice of the day-to-day worries of family life.

# Summary

Modern water management is characterized by an integrated approach of entire *water systems* and an increased concern for new interests that did not feature prominently in the past. Current policy objectives for water management focus on the creation and maintenance of a sustainable living environment, taking into account all demands made on the water system by the different interests. The intention is to consider a water system in its entirety. This implies that the relationships between various *subsystems* of a water system, such as surface water and groundwater in rural areas and surface water, groundwater and sewers in urban areas, should be considered together. These subsystems are traditionally the responsibility of different authorities.

Present-day planning processes involve balancing interests and setting priorities, scheduling the layout for the area concerned and combining or, conversely, splitting up interests. In day-to-day operational water management, this new development is still in its infancy. Various problems still have to be solved before the operational tasks of the water authorities of the various subsystems can actually be coordinated all the time. No impartial methods are available as yet for deciding which interactions between subsystems are really important and would therefore have to be incorporated in overall operational management.

Considering the aspects listed above, the main objective of the present study is: to develop a generally applicable methodology to achieve a well-balanced design and control of regional water systems, considering the dynamics of the intrinsic processes in the water system and the various requirements of the different interests, which may, in addition, vary in time.

A key aspect of the methodology developed here is a weighing mechanism that enables water managers to assign priorities to the various interests present a water system. Subjective policy preferences can be included in this weighing. The following types of interests are distinguished: common-good interests, sectoral interests and operational interests. Common-good interests involve requirements related to the primary water-management duties, such as flood prevention and maintaining sustainable conditions for the ecological obligations of water systems. Sectoral interests are characterized by the benefits that a particular group derives when specific requirements are met, or that are generally considered desirable in present-day society. Examples are: agriculture, recreation and nature preservation. Finally, operational interests consider efficiency in water management, such as the best possible water-system control at the lowest possible cost.

Demands made on water systems are time-dependent and in addition vary depending on the seasonal situation of the water system. For example: the requirements of arable farming on the water system are greatest during the sowing and growing season, of water sports in the holiday season and weekends, of navigation during transport on water and of nature when, for instance, wildflower seeds germinate in spring. To establish optimal control of regulating structures in all possible situations, the dynamics of the water system and the time-dependency of the requirements have to be taken into account. Typical for a *hydrological load* on a water system such as precipitation, is that it is dynamic and that it is difficult to predict exactly which quantity will fall at what moment. The dynamics of the water system, the demands it has to meet and the hydrological load, necessitate a dynamic approach to water management, which continually reflects the current situation. This type of approach is fundamentally different from current practice, where fixed target values in control of water systems are the main issues.

The type of approach that takes into account dynamic processes will here be called *dynamic control*. The outcome of dynamic control is a time-dependent *control strategy* that determines how the regulating structures of a water system can be applied best for a particular period ahead, to meet predetermined water-quantity and water-quality objectives as well as possible.

In day-to-day operations the different interests present in the various subsystems of a water system, rarely require the full *system capacity* simultaneously. Therefore, there is frequently room to employ the unused capacity of one subsystem for the benefit of another subsystem. A simple example is temporarily storing water in the subsurface to prevent flooding by surface water elsewhere.

If dynamic control is applied, a smaller overall capacity is required than would be expected on the basis of a static approach. The resulting excess system capacity can be used in two different ways. The capacity could either be used to better satisfy the growing number of requirements presented to the water system by newly recognized interests, or the excess capacity could compensate for any deficiencies in the system. In practice, the latter implies that planned extension of infrastructure can be postponed, reduced or canceled altogether. Examples are widening of water courses and/or building extra pumping capacity. This would yield considerable savings.

To enable the application of dynamic control, a real-time control system is required. A *Decision-Support System* (DSS) can help determine the best control strategy. The DSS especially developed during the present research is called AQUARIUS.

AQUARIUS consists of a number of interactive program modules: a *simulation* module, a *prediction* module and an *optimization* module. The simulation module accurately determines the current state of the water system, considering both water quality and water quantity.

The prediction module predicts the hydrological load to the water system for a particular period ahead, on the basis of weather forecasts. This period is here called the *control horizon*.

The optimization module builds the optimization problem for a period equal to the control horizon and resolves the optimal control strategy. The mathematical formulation of this problem comprises several intrinsic water-system relationships, the requirements presented to the water system by the various interests, the current state of the water system and the predicted hydrological load.

The optimal control strategy is transferred to the simulation module, which in turn determines the new state of the water system on the basis of the first control actions resulting from the control strategy. This technique, which involves program modules interacting and at the same time transferring data, is called *simultaneous simulation and optimization*.

The methodology incorporated in the DSS can be used for analysis purposes as well as for day-to-day operations. In the present study, the DSS has only been used for water-system analysis. In that analysis, multi-year time series were calculated and statistical information has been gathered from the calculation results.

While developing the DDS, the choice of an efficient optimization method and water-system modeling with its associated requirements were emphasized. The mathematical optimization method selected is *Successive Linear Programming* (SLP). The main justifications for this choice were: the speed of that method and the accuracy that can be achieved. The speed in finding the optimal solution of a control problem is particularly important in day-to-day operations.

The processes that take place in a water system are nonlinear. To ensure the required accuracy, nonlinear processes are linearized for the control horizon at the estimated values of the water-system variables. The method developed to do this efficiently is here called *forward estimating*. Calculating the time series involves building a optimization problem for each time step. In building the problem the forward estimate repeatedly uses the optimal solution of the previous time step.

In the case studies, the principle of forward estimating was found to be so efficient that the optimization problem only had to be solved once for each time step of the time series calculated. This makes the SLP method virtually as fast as standard Linear Programming (LP).

The method was found to enable determination within seconds of the optimal control strategy for a water system comprising dozens of rural and urban subsystems, where a limited number of requirements has to be met. Even for a large, complex water system, consisting of hundreds of subsystems, where a large number of requirements has to be met, the optimal control strategy can still be determined within minutes.

The reliability of weather forecasts supplied by the weather bureaus, was investigated. Using present-day techniques, predicting extreme precipitation events was found to be virtually impossible. A theory was developed to enable determination of a control strategy on the basis of the available forecast, which carries the least risk of violating the objectives of control that would result in e.g. floods or sewer overflows. An application of that theory is included in the prediction module of the DSS.

In the present research, dynamic and other forms of control were studied for water systems both for flat, polder areas and for gently sloping, hilly areas. The consequences of using dynamic control were analyzed in detail for the water systems in the areas controlled by the Water Boards of Delfland, De Drie Ambachten and Salland.

One of the parameters used in the evaluation of dynamic control is the *performance index*, which reflects how well the requirements set for a water system are met over an extended period, when a particular control mode is used. Time-series calculations were used to compute the performance indices. In addition, the consequences of using dynamic and other forms of control have been evaluated by checking *system-failure frequencies*, *failure duration* and several other performance parameters. Extreme situations, such as high water levels and extremely poor water quality have been investigated.

For both polder and hilly areas, dynamic control was found to yield a considerable improvement in meeting the objectives set for water systems. In general, the improvement in performance that can be attained by using dynamic control is largest in flat polder areas, because these water systems are relatively easy to control. In hilly areas, where the soils are often sandy, a considerable amount of precipitation may percolate via the groundwater, sometimes even resulting in water courses running dry. At present, the controllability of water systems is usually limited in these hilly areas. However, this situation would change if the water management system of these areas is adapted to the requirements, which may include the construction of water-supply works.

Furthermore, the susceptibility of the water systems to extreme events such as massive downpours or droughts, was analyzed in the case studies. The general conclusion is warranted that dynamic water-system control more efficiently uses the capacities that exist within the water system, for instance the surface-water storage capacity, the groundwater storage capacity, the sewer-system capacity and the pumping or inlet capacity. An accurate prediction of the hydrological load was found not to be essential in all cases. Especially in water systems where discharge is slow, such a prediction may not be necessary at all. However, in fast-discharging systems, such as sewer systems, a more accurate prediction is generally required.

The analyses performed using dynamic control prove that this control mode satisfies the current requirements in water systems better. Moreover, the flexibility resulting from dynamic control in operational management often yields excess system capacity. The excess capacity available is sometimes sufficient to compensate scheduled extensions in a water system without having to build new infrastructure nor enlarge the existing structures.

A prerequisite for dynamic control, however, is the availability of a real-time version of the DSS, situated in a central location, where monitoring data on water quantity and water quality are available. Depending on which mode of operation is preferred, either *dynamic automatic control* using control units, or *dynamic manual control* by operators can be selected.

The prospect of efficient water-system control and the resultant optimal use of existing system capacities, raises the question whether application of this methodology has to be taken into account in the design stage of new water systems. After all, unnecessary and unusable excess system capacity might be avoided. To test this option, a *dynamic design procedure* has been developed.

Standard design methods are based on generalized rules and use generalized data on soil properties and use. The dynamic design procedure also takes into account the layout of the water system and the actual conditions of soils, waterways, and regulating structures that exist in a water system. In addition, stochastic methods assist in assessing the operational situation of the control system, including those situations when parts of the system are unavailable as a result of maintenance or technical malfunctioning.

The dynamic design procedure is based on verification of the performance of a particular water-system design. The merits of a design are verified using time-series calculations that determine the *failure frequency* and *failure duration* of the water system. Failure in this respect is defined as the situation that arises when one of the requirements identified for the various interests, is violated.

When verification is applied to systems that were designed using traditional, standard design methods, the associated design rules were found to result in water systems that have unnecessary large excess capacities. However, the exact sizes and locations of these excess capacities cannot be determined from the standard design.

The dynamic design procedure takes into account the actual situation of the water system and determines the system capacities exactly needed for water management, whether dynamic control is really applied or not. The method results in a well-balanced water-system design, where sizes and locations of any excess capacities are known exactly and which, obviously, satisfies the requirements identified, including the obligatory safety margins.

Last but not least, water systems, in which dynamic control was incorporated at the design stage, were found to be much more cost-effective to build than those intending to use local automatic control. The conclusion is therefore warranted that, on economic grounds, the intended control technique should be taken into account as early as the design stage of a water system.

A.H. Lobbrecht (1997). Dynamic Water-System Control; Design and Operation of Regional Water-Resources Systems. A.A. Balkema, Rotterdam, NL.

*Fig. 1.1. Location map.*

# 1 Introduction

## 1.1 Framework

### 1.1.1 Scope of Research

A major recent development in the water sector is the gradual introduction of integrated water management in the Netherlands and several other countries. Present-day integrated water management presents a way of considering the entire water-related environment, in which various interests are present, each of which poses its own specific demands. An essential element of integrated water management is the *water-system approach*, which considers the various interrelated elements of a water system and their interactions.

A problem that crops up is that the number of interests considered important increases continuously, whereas the water systems cannot always meet the corresponding demands. Considering the great number of water-related interests, weighing is needed. This is achieved in the planning stage of water systems by setting priorities and ranking.

A major issue in this thesis is the allocation of scarce or plentiful water resources to the various interests (Fig. 1.2) at the right time. In that allocation, the requirements of the interests should be met as well as possible, taking into account both water-quantity and water-quality aspects.

Many interests that are currently considered important, use the same water resources and are geographically linked. A tendency can be observed to separate interests that have conflicting

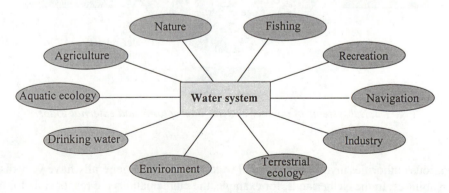

*Fig. 1.2. Water-related interests.*

water-quantity and/or water-quality requirements. Separating activities and relocating them to different sites in the environment can result in a drastic change in historically grown situations. Such relocations are no doubt very costly.

A possible effective alternative is to leave the environment unaltered as much as possible and adapt the management and control of the water system in such a way that the requirements of each interest are met adequately. This approach can be very flexible as it involves control by a better use of often existing infrastructure. The control of a water system can thus be adapted to changed circumstances with relative ease.

In general, very little attention has been given to the consequences of integrated water management schemes that involve the design and operation of an entire *water system*. Most traditional methods for design and operation are based on quantitative standards. Such standards are typically fixed for many years. The standards incorporate extreme conditions in the water system in the design, whereas in day-to-day operations, the average conditions are generally considered more important.

After a long history of traditional methods to control *subsystems* (Fig. 1.3), water authorities are currently adopting more integrated control methods. However, the increasing number of demands results in a complex operational problem, especially during periods of extreme hydrological conditions, such as excessive rainfall or severe drought.

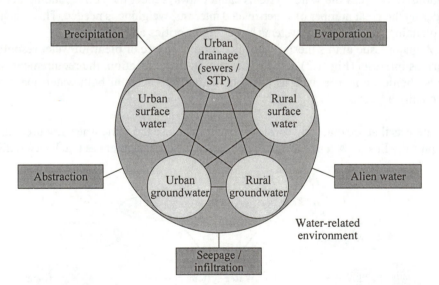

*Fig. 1.3. Subsystems of a water system: interactions and external loads.*

The various authorities involved in regional water management generally have very distinct responsibilities. In the Netherlands, for example, the municipalities are responsible for urban drainage, the water boards for surface-water quantity, surface-water quality and sewage treatment, and the provinces for regional groundwater.

Objectives for control in one subsystem may be in conflict with objectives in another subsystem. The strict administrative borders in operational control and the current practice of trying to meet the objectives of each subsystem separately, preclude a successful balance of all interest requirements within a water system.

To achieve integrated water-system control, water authorities should therefore overcome the administrative hurdles and aim for real integration, not only in the planning stage, but also in the design and operation of water systems.

## 1.1.2 Objective of this Thesis

The present study presents a methodology for the operation of water systems on the basis of the definition and weighing of time-varying interests. Existing water-system arrangements are taken as a starting point to determine control strategies that best satisfy the requirements of each interest. Another issue of importance in the present study is to extend the interest-based approach to water-system design, to include the operational situation. This incorporates the necessary control-system and the water-system dynamics.

The thesis describes water systems on a regional scale, involving a wide range of interests. The purpose is to fill the gap between the current integrated water-management policies and their implementation in the operational setting.

Special attention is given to the various types of regional water systems that can be distinguished. The main focus is on rural areas, but urban areas including sewer systems are incorporated as well. Large river basins have not been considered in the scope of the present study.

The challenge of the study is to solve the operational water-control problem that occurs when trying to meet the objectives of various and sometimes conflicting interests present in the water-related environment. General examples of such objectives are:
- to prevent flooding in rural and urban areas;
- to maintain conditions that enable sustainable flora and fauna;
- to maintain biodiversity;
- to prevent drought problems in rural and urban areas;
- to facilitate transport on water;
- to prevent poor surface- and groundwater quality;
- to minimize operational costs.

Summarizing, the objective of this thesis is to develop a generally applicable methodology to achieve a well balanced design and control of regional water-systems, considering the dynamics of the intrinsic processes in the water system and the various requirements of the different interests, which may, in addition, vary in time.

### 1.1.3 Outline of the Thesis and Conventions Used

The first chapter of this thesis describes the problems and developments in current water management and places these developments in a historical perspective. It furthermore gives a general introduction to the subject of dynamic control.

Chapter 2 shows the general structure of a control system and introduces the specific terminology used with respect to the control of water systems. Special attention is given to Decision Support Systems and how these systems can be applied in operational water control. Chapter 3 mathematically formulates the general operational control problem and outlines various methods for solving that problem. Chapter 4 describes the various elements of regional water systems that are used in modeling. Chapter 5 presents an analysis of hydrological loads on a water system and the way in which these loads can be determined on the basis of weather forecasts and other prediction methods. Chapter 6 explains the way in which the control problem considered is formulated, specifically focusing on efficient solving. Special techniques are presented to keep the solution to the control problem accurate. Chapter 7 describes applications of the method developed in several case studies. Chapter 8 focuses on the consequences of control to the design of water systems. In Chapter 9 the overall results are presented, as well as conclusions and recommendations for further research.

The reader who is interested in the subject of dynamic water-system control, but not specifically in the underlying methods, is recommended to read only the current chapter and Chapter 9.

All through this thesis keywords are defined. Keywords that are used frequently are printed in italics and are included in the attached glossary. Other keywords that only have a specific meaning once or twice in the text, are printed between quotation marks. Abbreviations are explained when they are used for the first time. All abbreviations are included in the list of abbreviations.

The relevant variables used in this thesis are summarized in the attached list of symbols. Multiple use of the same variable name with a different meaning is avoided as much as possible. Some equations have a very specific naming convention though. In those cases it happens occasionally that a variable is used for the second time with another meaning. However, this is explicitly indicated in the text and these variables are not incorporated in the list of symbols.

## 1.2    Water-System Control in Historical Perspective

### 1.2.1  General

This section outlines the historical developments that resulted in present-day regional water systems in the Netherlands and the way in which they are controlled. It describes

*Fig. 1.4. Typical water-management regions in the Netherlands.*

developments in water-system control and tries to generate an understanding of current water-management problems and the way in which these problems can be solved. It is not the purpose to discuss the history of water management in the Netherlands in detail. Comprehensive descriptions of the reclamation of land which previously belonged to the sea can be found in Colenbrander (1989), Schultz (1992) and Van de Ven (1993). The historical overview described in this section is based on these books.

The present-day Netherlands can be roughly divided into a flat and low-lying region, the majority of which is below mean sea level (MSL) and a more hilly region, the highest parts of which reach up to 300 m above MSL (Fig. 1.4). The low-lying region consists mostly of polders, drained by pumping stations. For that reason the areas in this region are here called *polder areas*. Areas in the hilly region, which in most cases are drained by gravity, will here be referred to as *hilly areas*. It should be mentioned that these hilly areas, in comparison to the hilly areas in some other countries are still rather flat.

### 1.2.2  Polder Areas

The majority of the polder areas is located in the north-western part of the Netherlands and along the main rivers Rhine and Meuse. Reclamation of these areas started around the year 800 AD. Many of these areas were frequently flooded by sea or river water and therefore consisted of swamps, peat bogs and lakes. The average altitude of the areas along the coasts was equal to mean sea level during that time. Reclamation, mainly for agricultural purposes, such as arable farming and cattle farming, took place by means of simple drainage works and small local dike construction. Openings in the dikes were used to discharge excess water that logged the lands.

Most of the present-day polder areas were originally covered by peat and clay on peat. Reclamation of these areas lowered the groundwater table, resulting in subsidence of the soil surface. The main processes that caused this subsidence were the increased soil pressure and oxidation of peat. Moreover, the peat itself was excavated by the inhabitants of these areas for salt and fuel production. Both effects, land reclamation and peat excavation resulted in a lowering of the surface by one to two meters. The lowering allowed tidal water to enter more easily via creeks and rivers, which was one of the reasons of the severe floods that occurred between 800 and 1250 AD. Towards the end of his period, dike construction improved, creeks were dammed and simple spill sluices were built. These sluices were used to discharge excess water during low tide and prevent inflow during high tide. The height of dikes had to be adjusted frequently, as each new extreme water level determined the new required height. Nevertheless, many floods occurred.

Between 1250 and 1600 AD many of the present-day regional water boards in polder areas were established as water-management agencies. Their main task was to manage the main dikes around polders. In some areas, polder water boards were formed to take care of inner dike rings, drainage works and discharge structures. The continuing soil subsidence in the coastal regions, the sea-level rise and the rise of the river beds of the major rivers reached such an extent that simple gravity discharge to sea and river water was no longer possible in many regions. Therefore, around the year 1400 AD, the first windmills, which were previously used for milling corn, were converted to discharge polder water. Paddle wheels could lift excess water by a maximum of 1.5 m. Discharging water outside the polder dikes created *storage basins* ('boezems') between the polders. Storage basins were, and still are, used to temporarily store polder water and discharge it gradually to the sea or rivers. These storage basins were originally discharged by gravity. However, especially during high water levels of the rivers or the sea in winter, the discharge capacity of the storage basins was not sufficient. For that reason, a *milling stop* ('maalstop') could be imposed by the regional water board, which meant that some or all windmills had to stop discharging into the storage basin to prevent polder dikes overflowing. Therefore, during long periods and sometimes during an entire winter, the polders could not be drained and were sometimes entirely flooded as a result of a milling stop.

Advances in windmill and dike construction led to improvements in water management around 1600 AD. Large elevation heights became possible by placing windmills in series. This development allowed the lower-lying lakes to be drained and reclaimed. These areas are known as *reclaimed lakes* ('droogmakerijen').

In the nineteenth century mechanized pumping by steam engines was introduced, so drainage no longer depended on the wind. The Haarlemmermeer polder was the first polder to be drained by steam-driven pumps. New types of machinery such as the centrifugal pump were introduced, providing an enormous increase in pumping capacity and elevation height in comparison to windmills.

In the lower regions of the Netherlands, where water levels are far below that of the surrounding land and water levels, reclamation introduced the problem of brackish to saline seepage from deep groundwater.

In addition to the polders, the storage basins were also drained by means of steam-driven pumping stations. This meant a real improvement in water management because now water levels could be maintained during the entire year.

Around 1900, diesel engines were developed and were soon applied for polder and storage-basin drainage. Soon after, pumping by means of electrical engines was introduced. The use of diesel and electrical engines enabled construction of large pumping stations with high capacities. Using pumping stations like these, parts of the largest lake of the Netherlands, the IJsselmeer, a former inland sea, were reclaimed: the Wieringermeer (1930), the Noordoostpolder (1942), and the Flevopolders (1957/1968).

Since the beginning of the twentieth century both diesel and electrical pumping stations were built. In practice, the construction of diesel pumping stations is more expensive, whereas

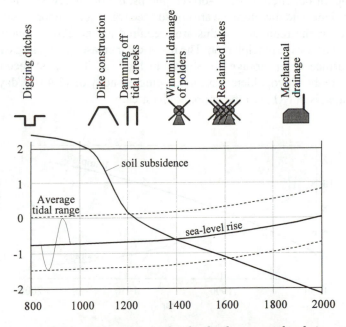

*Fig. 1.5. Schematic overview of soil subsidence, sea-level rise and technical advances between 800 and 2000 AD (after Luijendijk & Sinke, 1982).*

operation is less expensive than that of electrical pumping stations. The former 'Act for the protection of water structures in times of war' stipulated that for safety reasons a certain amount of drainage capacity in the polders should be provided by diesel-driven engines. For that reason, approximately half the number of large pumping stations is still of the diesel-driven type.

Figure 1.5 gives a general impression of historical events in polder development in the western part of the Netherlands and the consequences of subsidence of the soil surface and sea-level rise. The picture is representative for the majority of previously peat-covered polders in the Netherlands. The figure clearly shows the period of subsidence, caused by drainage and peat reclamation. Moreover, mean sea level slowly rose during the course of time. The graph shows that from the beginning of this millennium, land reclamation and sea-level rise increased the need for artificial control of water systems.

The majority of polders are not situated along the coasts, but more inland. These polders drain to river water. Since the river bed has been rising during the same period as indicated in Fig. 1.5, drainage of the more inland polders experienced the same type of problems.

The capability to control the water level in a polder system depends mainly on the design variables 'pumping capacity' and 'storage capacity'. In the times of windmill drainage, water-level control relied on large bodies of open water for temporary storage of excess water. Once the wind picked up, the mills could discharge the excess water.

After the introduction of steam-driven pumps, the discharge of water from polders became more reliable and therefore the area of surface water for temporary storage could be reduced. However, the reduction of this area required an increase in pumping capacity, especially during excessive precipitation. The historical increase in pumping capacity and the reduction in surface-water storage are shown in Fig. 1.6. The most recently reclaimed polders, which are the Flevopolders, have a pumping capacity of 13.4 mm/day, whereas the surface-water area is only 1% of the entire polder area.

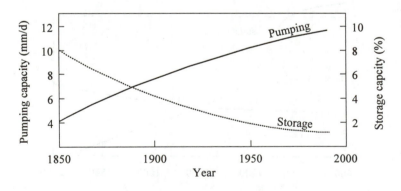

*Fig. 1.6. Historical trends in required pumping and storage capacity*
*(free after Schultz, 1992).*

## 1.2.3 Hilly Areas

The topic of this section is the development of water management during the past few centuries in the hilly areas of the Netherlands. These areas are situated in the east and south of the country (Fig. 1.4). The hilly areas are generally gently sloping and the crests reach up to 50 m above MSL. An exception is the very southern part of Limburg (Fig. 1.1) where hills exist of up to 300 m above MSL. The discussion in this section is restricted to gently sloping hills.

In general, the soils in hilly areas are sandy, but peat soils exist as well. These areas used to be drained naturally by gravity through small rivers and creeks. Rivers and creeks used to flood the adjacent lands frequently, especially in winter.

Originally, most areas with sandy soils were covered by forests, only the highest areas have always been almost bare. Large parts of these forests have been cut down for extensive agricultural and other purposes. Therefore, large areas covered with heaths developed. Reclamation of these areas started by the end of the nineteenth century. During that time deciduous and evergreen trees were planted in the higher sandy regions for several reasons: to increase evaporation and thus lower the groundwater table; for timber production and to bind drifting sand dunes. Canalization of creeks enabled a faster discharge of water. The increased discharge, in some locations in combination with the construction of dams for milling purposes, resulted in floods along the lower reaches of the local rivers and creeks. Subsequently, these had to be trained and embanked to prevent these undesirable situations.

Peat excavation for fuel production started during the nineteenth century. Canal systems were constructed to drain the areas and transport men and peat by ships. The canals discharged excess water into the main rivers or directly into the sea in the north of the Netherlands. As a result of peat excavation, the water-storage capacity of the soil decreased, which caused increased drainage and peak river discharges. At the end of the nineteenth century, water boards were established in the hilly areas to supervise water management.

Improving drainage and preventing floods have traditionally been the major concerns of water boards in hilly areas. Rapid discharge of water has been accomplished by continuous canal-profile enlargement, river lining and shortening the courses of rivers and creeks, by cutting off bends. All these activities resulted in lowering of the groundwater table, but also in creeks and rivers having a reduced base flow, high peak flows with high velocities in winter and often dry beds in summer.

To remedy this undesirable situation, water boards constructed weirs in all types of canals, rivers and creeks during the past few decades. Especially in areas where soils consist of coarse sand, the results of these measures were poor. As a result of surface-water runoff and shallow seepage below the structures, discharge was still too fast (Fig. 1.7). At present, a rate of one weir per 200 hectares is not uncommon, which means that water boards in hilly areas have to supervise and maintain a large number of weirs.

The improved conditions for agricultural activities in hilly areas, the use of intensive drainage and the resulting lowering of the groundwater table, have had a serious impact on scenery, nature, flora and fauna. Previously swampy areas flooded frequently, became dry land, where first forests had been replaced by heaths and at some locations heaths had again been replaced by arable land and pastures.

Traditionally, water is abstracted from the soils in sandy hilly areas for drinking-water production. Obviously, these locations were originally selected because of the good quality of the groundwater, which, in principle, did not need treatment and furthermore, the good storage capacity of the soil. However, groundwater abstraction for these purposes further accelerated the lowering of the groundwater table.

The past few decades, a shortage of water has developed in hilly areas because the groundwater table has been lowered too much. In some areas the effects of lowering the groundwater table have had disastrous effects on the natural vegetation and agricultural production.

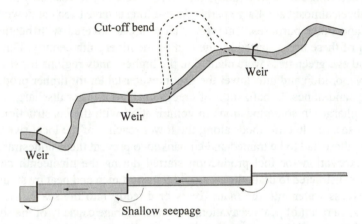

Fig. 1.7. Surface-water runoff and shallow seepage in hilly areas where the soil is very permeable (plan view and longitudinal cross-section of a canal).

### 1.2.4 Current Situation

*Land Use*

The surface area of the Netherlands at present covers approximately 30,000 square kilometers, about half of which are polder areas and half hilly areas. The current population is around 16 million. The most densely populated region, in the western and central part of the country, commonly called the 'Randstad conurbation', consists mainly of polder areas. Most of the rural areas are used for agricultural purposes, where people live in a large number of relatively small villages.

The present-day landscape of the Netherlands is determined by the historical battle against sea and river floods. Villages were to a large extent built on high grounds such as river levees and dikes. Long stretches of ribbon building are very common in these areas. Access to these areas is often by roads constructed on top of dikes.

New urban developments mainly take place in polder areas and in the lower lying lands along rivers. Extensive forms of agriculture such as arable farming are slowly decreasing and agricultural land is being converted into urban areas and nature reserves. Conversely, intensive forms of agriculture are increasing, especially horticulture in glasshouses and bulb farming.

With the increase in urban areas, the total impervious surface area consisting of roofs, roads and parking areas also increases. These developments affect runoff characteristics of the water systems and the way in which they have to be controlled.

In the polder areas, arable farming and cattle farming are the main activities. Pastures are found predominantly in polders, especially in areas where the groundwater table is shallow.

In the hilly areas, mixed forms of agriculture are found, mostly arable farming and cattle farming. Nature reserves are mainly found in the eastern and southern parts of the country. Here, dunes, heaths and forest are present. Cultivation of the hilly areas and development of residential areas has strongly affected the unspoiled natural landscape at many locations.

It is government policy to reconstruct the natural landscape in specific zones throughout the country. This includes formation of nature zones that are linked with each other, thus forming a natural belt that allows flora and fauna to develop and spread over a considerable surface. The idea is that certain species should again be able to migrate through an ecological route, unhampered by intensive agricultural areas, roads and urban areas.

Regional water control is the responsibility of approximately 65 water boards at the moment. This number has decreased considerably over the past few years because small water boards, especially those in hilly areas, have merged into new and larger organizations.

The main responsibility of water boards was traditionally surface-water-level control. These days, this task is broadened to include rural groundwater control, water-quality control and management of nature reserves. The complexity of the operational control task of water boards thus increased.

### Water Quality

Water-quality control is considered an important issue in both polder and hilly areas. The general objective is to maintain the water-quality standards specified for the interests present in a water system. Water-quality control can best be accomplished by means of controlling the source. This is, unfortunately, not always possible. For example, in low-lying polders, upward seepage of saline water is not prevented easily.

Several types of pollutants can be distinguished, some of which are related to the use of the water system, e.g.: excess fertilizers from agricultural areas, toxic substances from timber canal linings, effluents from sewage treatment plants (STPs), combined sewer overflows (CSOs) from sewer systems. Reducing the quantity of these sources of pollution is one of the main concerns of the water boards.

In a dry summer, surface water may consist for the larger part of the effluent of STPs, which can lead to very high nutrient loads. CSOs may occur several times a year, which causes a sudden and high pollutant load. If necessary, such pollutants are flushed from the surface water by importing relatively good-quality water from outside the polluted area.

Water from outside the water system carries other organic and anorganic substances than those specific for that system. Letting in such *alien water* is considered an important reason for the disappearance of location-specific species and the introduction of general forms of flora and fauna. Therefore, the current water-management practice is to avoid this inflow by preserving area-specific water during winter and spring as much as possible.

### *Drought Problems*

In summer, water shortages occur at many locations in the country. Preservation methods are being developed to keep as much water as possible within polder and hilly areas and prevent unnecessary outflow or inlet of alien water. In general, the water authority of a polder area can control water quantities discharged and let in relatively well, which allows accurate preservation. In hilly areas, however, special measures are required to enable water preservation. In general, two methods are distinguished: outflow reduction and water retention.

The principle of outflow reduction is that, in spring and in summer, surface and groundwater levels are kept the highest possible, to reduce outflow from groundwater and surface-water subsystems. This is currently done by reconstructing river bends and reduction of flow profiles, but also by control systems, such as weirs that can be lowered automatically in case of increased runoff and be raised in case of small or no runoff. Another form of outflow reduction, which is widely practiced in new urban areas these days, is connecting impervious surfaces to infiltration sites instead of to sewer systems.

The second method involves water preservation, using artificial retention reservoirs. It is applied in a growing number of cases. Especially in the hilly area of Limburg (see Fig. 1.1) such measures are taken in combination with downstream flood protection along rivers and creeks. Water-retention reservoirs can also be used to infiltrate river runoff and recharge the groundwater system. However, the effectiveness of such recharge very much depends on the permeability of the subsurface. In some polder areas this method is practiced as well, for instance, to temporarily store water which cannot be discharged to a storage basin.

Water shortage as a result of lowered groundwater tables is a major problem in the water management of hilly areas. Despite various efforts to prevent the inlet of alien water, the current practice is to compensate for water shortages by importing water from outside the region and artificial irrigation of agricultural land with sprinkler systems. Water supply works are carried out in many parts of the hilly areas. Water is brought in by canals and small pumping stations. Sometimes existing drainage canals are used for this purpose, reverting the direction of flow.

The IJsselmeer and the rivers IJssel, Rhine and Meuse are the main sources for irrigation water in the Netherlands (Fig. 1.4). To prevent flooding along supply canals, water is generally pumped up in small steps. For sprinkler irrigation, surface water and groundwater are used, if available. The use of groundwater by farmers for these purposes is restricted in dry periods by the responsible water authorities, to prevent excessive lowering of the groundwater table.

## 1.3   Developments in Water-System Control

### 1.3.1   Lack of Capacity

Designing and operating water systems is a matter of capacity allocation. The factual reason that water levels rise and for the subsequent flooding or overflow, is a lack of storage or discharge capacity in subsystems, such as surface-water, groundwater and sewers systems. Undesirable situations in these cases can be prevented by larger storage or discharge capacities. Despite this fact, undesirable situations continue to occur, since the capacity for storing water in and abstracting from a subsystem is limited in practice. The fact that a specific situation is undesirable, follows from the requirements of interests defined for subsystems. An undesirable situation occurs if one or more of these requirements are violated. That situation will here be called *system failure*. The frequency of system failure used in designs of water systems depends on the social and economic acceptance the effects.

In general, the frequency of system failure can be reduced by an increase in system capacities. However, other possibilities can also contribute to such reductions. Research shows that usually not all the capacity available is used at the moment of failure and that unused capacities remain in the system (Schilling, 1991). Sewer-system engineering has shown that system failures can be limited considerably by installing control systems, which reallocate available storage and discharge capacities before the moment of failure. To accomplish such subsystem control, the available capacities have to be used effectively in time. Several control systems that apply this have been implemented for sewer systems.

Considering the results achieved for sewer-system control, it is of interest to know whether similar results can be obtained in the control of entire regional water systems. Subsystems of a water system rarely fail at the same moment. Therefore, the temporarily unused capacity of one subsystem can be used in favor of another subsystem.

Because of the growing number of interests, the number of requirements that have to be met in water systems increases considerably. Therefore, the freedom of control is becoming smaller. Part of this problem can be solved by additionally making use of the variability of interest requirements throughout the year. For example, if an agricultural interest requires constant high groundwater levels only during the sowing and growing periods, groundwater levels could be kept low outside these periods, using the groundwater-storage capacity to reduce the chance of flooding in an other location.

The traditional way of controlling water systems is to determine fixed target levels and control ranges as standards. Under average conditions, which occur in theory only, the requirements of all interests are satisfied best using these standards.

In practice, however, various interests with conflicting requirements can be present in a water system. The operational control problems that occur as a result of the increase in importance of interests, which were not considered previously, necessitate using all the available system capacities.

### 1.3.2  The Operator's Role

The traditional manner of controlling subsystems of a water system is shown in diagram (a) of Fig. 1.8. The operator has a central position. He receives standards set by policy makers. The operator gathers information about the functioning of the subsystem from observation. On the basis of this information and the standards, the operator controls the processes which take place in the subsystem he has to control. The operator monitors the results of his control actions and the behavior of the water system. This information and his experience enable him to determine how the system should be operated to continuously meet the standards. Because changes take place in the water system and standards are updated regularly, the operator has a limited time window for his experience to grow.

One of the drawbacks of this traditional form of control is that each operator is responsible for one subsystem only. Therefore, he tends to focus on the performance of his own subsystem. In practice, for a long period this has resulted in accurate control of surface-water levels in regional water systems, whereas groundwater levels and water quality received only limited attention.

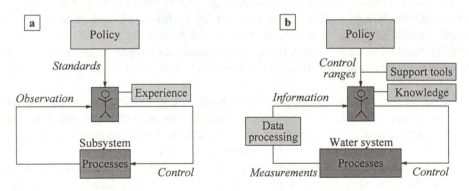

*Fig. 1.8. Traditional (a) versus modern (b) control of a water system.*

In modern control, several subsystems and preferably the entire water system is involved (diagram (b) in Fig. 1.8). Information on both water-quantity and water-quality variables are monitored continuously or at regular intervals. These data are gathered from various locations that are geographically apart and processed into suitable information. In this concept, the operator receives more information about the behavior of the water system and the processes involved. He has to deal with more standards than when controlling one subsystem only. The requirements of various interests that have to be satisfied in a water system are presented to the operator in the form of control ranges. These allow him more flexibility of control. To support his decisions, the operator has access to tools that enable him to evaluate various control options in advance. Such tools are, for example, models of the water system that incorporate an accurate representation of the water system and the control structures present.

In this modern control system the operator gains experience about the functioning of the water system. On the one hand, he has more data to base his decisions on, but on the other hand, his support tools enable him to gain knowledge much faster than before. This is because he can 'play back' earlier system loads that resulted in extreme conditions and necessitated difficult decisions.

Another variable which can be of help to the operator is a weather forecast, which can be translated into a prediction of the water-system load. This enables him to anticipate runoff in case of rainfall or drought. Monitoring and system-load prediction even enable him to take control actions before an event actually takes place.

### 1.3.3  Automation of Control Systems

Water authorities are currently changing over from traditional forms of manual control to various kinds of automatic control. Automation often means an increase in efficiency, but not necessarily improved control. To really enhance the performance of water systems, automation should be introduced to execute operations that were not done before. In general, water authorities start with *on-line monitoring* which includes automatic data gathering of, for instance, water levels and flows.

A survey among regional water authorities showed that almost all newly installed water-quantity monitoring devices function on line (Lobbrecht et al., 1995). An important reason for this development is the need for a more accurate picture of the water-system state, which can be used for operational water-management and planning purposes.

Figure 1.9 shows the results of a survey of approximately 8000 regulating structures, of which 900 were polder and storage-basin pumping stations. The figure shows that pumping stations are automated to a large extent.

Pumping stations of polder areas that were previously attended and operated manually during the day only, are now being automated. The continuous availability of pumping capacity enables a more accurate surface-water-level control. At any moment, even during the night, pumps can be switched on if necessary, while operators do not need to be present. This is clearly demonstrated in Fig. 1.10, which shows that the surface-water level in the Flevopolders, is fluctuating within a much narrower range after automation of two if its pumping stations. The data are from two hydrologically comparable years, 1987 (before automation) and 1991 (after automation).

In hilly areas, weirs are being automated to maintain the highest surface water levels possible and preserve as much groundwater as possible in summer. Throughout the year, the automation systems control up- or down-stream water levels. During excessive rainfall, weirs are lowered automatically, whereas during periods of drought they are lifted.

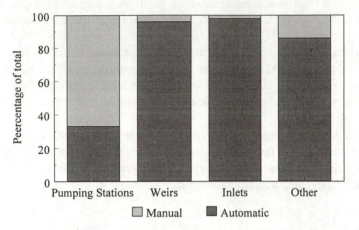

*Fig. 1.9. Current extent of automated regulating structures (from: Lobbrecht et al., 1995).*

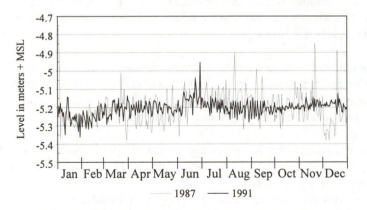

*Fig. 1.10. Water levels in the High Section of the Flevopolders, before (1987) and after (1991) automation of pumping stations (from: Verbrugge & Lobbrecht, 1993).*

### 1.3.4 The Evolution of Water-System Control

Automation of routine tasks is one of the first steps towards improved water-system control. However, automation of routine tasks alone is not sufficient to meet many of the present requirements. The awareness is growing that a weighed form of water-system control is necessary, in which automation plays a role.

Three evolutionary steps can be distinguished that will eventually lead to weighed control of a water-system (Fig. 1.11):
1.　local control,
2.　central control,
3.　dynamic control.

Local control involves a single regulating structure in a water system and is executed on the basis of monitoring data gathered in the vicinity of that structure. Local control takes place on the basis of standards that have been set for each subsystem. This form of control is practiced in many water systems by pumping stations that control surface-water levels and weirs that control upstream water levels.

Central control involves one or more regulating structures and is executed on the basis of data from more than one location in the water system. Several subsystems can be involved in central control. Similar to local control, central control takes place on the basis of pre-set standards. The advantage over local control is that, because a better picture is available of the water-system, the required water-system state can be determined, avoiding unnecessary or contradictory local control actions. Central control is currently implemented by several water-management agencies by using data from monitoring networks. A central control mechanism generally implies logic control rules. An example of this type of control is the combination of surface-water-level and water-quality control.

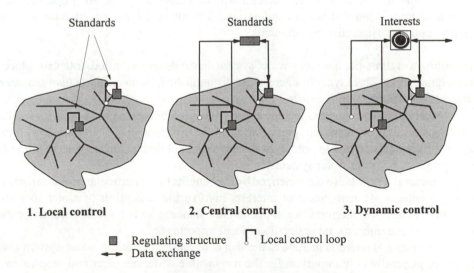

*Fig. 1.11. Three steps in the evolution to dynamic control of a water system.*

Dynamic control is a specific mode of central control, in which control actions are based on the time-varying requirements of interests in a water system, the water-system load and the dynamic processes in the water system. Dynamic control incorporates a mechanism that enables continuous weighing of interests present in the various subsystems. Using dynamic control, the available system capacities can be used optimally both under normal and exceptional conditions. During extreme hydrological circumstances, subsystems of the water system may fail. Using dynamic control, failure in each subsystem is avoided as much as possible, given the available water-system capacities. The weighing mechanism used, ensures that if failure cannot be prevented, the least important interests fail first and the most important ones last.

It should be mentioned that the steps presented in Fig. 1.11 rarely exist in pure form. A mix of local and central control is already found in various water systems. Dynamic control is not yet applied in the Netherlands. It is expected that dynamic control will be introduced gradually. Especially in the beginning, it will be implemented mixed with central and local forms of control.

It should be stressed that central and dynamic control do not necessarily replace the control actions of operators. Both forms of control can be useful in the practical operational situation using manual operations. For instance, large pumping stations can be operated manually, while decisions are made on the basis of the outcomes of central logic.

## 1.3.5 Interest Weighing

In the Netherlands, 'functions' are assigned to specific locations in the environment by the government. A function in this respect is the formalized main interest. The definition of a function includes water-quality standards, which determine the environmental conditions that have to be satisfied.

In this thesis, only those interests will be considered that are important for the determination of control strategies and designs. The interest definition includes both water quality- and water-quantity requirements.

According to a three-tier hierarchy water-system demands are classified: interests, objectives and requirements. Three types of interests are distinguished: *common-good interests*, *sectoral interests* and *operational interests*.

Common-good interests include aspects such as whether the land is fit for habitation and focus on improvement and preservation of the environment. Common-good interests entail requirements related to primary water-management duties. These types of interests are of importance to entire water systems.

Sectoral interests are characterized by the benefits that a particular group derives when its requirements are met. Sectoral interests involve the allocation of water to different activities. In sectoral interests, socio-economic balancing and/or social acceptance can be involved. These interests are generally of local importance.

Operational interests include issues that determine an efficient water-system control. They are generally only important for the responsible water manager and require the best operational control at the lowest cost.

Table 1.1. Examples of interests and possible associated objectives and requirements.

| Interest | Objective | Requirement |
|---|---|---|
| Flood prevention (common-good) | Prevent flooding in rural and urban areas | Keep water levels in rural and urban systems below the set limits<br>Prevent scouring of canal beds and embankments |
| Ecology (common-good) | Prevent drought<br>Prevent pollutant influx into surface and groundwater<br>Maintain biodiversity | Maintain surface- and groundwater levels<br>Prevent undesirable water-system inflows<br>Preserve location-specific water<br>Use sewer storage optimally before overflow |
| Agriculture (sectoral) | Maintain optimal growing conditions for crops<br>Keep the land accessible | Keep groundwater levels within the set range<br>Keep soil moisture contents within the set range<br>Maintain water in the surface-water subsystem for irrigation<br>Keep surface-water- and groundwater-quality variables within the set limits |
| Navigation (sectoral) | Facilitate transport on water | Keep surface-water levels within the set range<br>Keep flow velocities within the set range |
| Water recreation (sectoral) | Maintain good surface-water quantity and quality | Keep surface-water levels within the set range<br>Keep surface-water-quality variables within the set limits |
| Nature (sectoral) | Maintain sustainable conditions for flora and fauna | Keep groundwater levels within the set range<br>Keep surface-water levels within the set range<br>Keep surface- and groundwater-quality variables within the set limits |
| Water management (operational) | Minimize operational costs | Minimize the number and duration of operations<br>Keep flow velocities within the set range |

Objectives set for interests, express the preferred situations in the water-system. The requirements of interests represent the intrinsic water-system conditions to be met.

It should be kept in mind that in this thesis the main focus is operational control and not determining the layout of water systems. Therefore, interests are presented and dealt with mainly from an operational point of view. The objectives of various interests may be the same, but are generally different. Table 1.1 presents some examples of interests distinguished in present-day water-system control, with possible associated objectives and requirements. The order in which the examples are given does not represent a ranking, the table is only for demonstration purposes.

The weighing mechanism used in dynamic control enables interests of different kinds to be compared. Dynamic control determines which interests are preferred above others by means of the assigning of weights. During normal operating conditions, the resulting control may be very similar to central forms of control or even to local control. Under exceptional conditions, for instance, during excessive precipitation or drought, the actions of dynamic control shift in favor of interests that have been assigned the largest weights. However, average conditions may also require dynamic control to optimally meet the requirements of the interests considered.

## 1.4    Decision Making

This section briefly introduces the decision-making process as currently practiced by water authorities such as water boards. The terminology used for the various organizational levels in the next chapters of this thesis is introduced. The only purpose is to present the organizational framework and how decisions are made and not to discuss the procedures of the decision-making process itself.

Two terms are used to indicate the organization responsible for water management: *water authority* and *water manager*. The term water authority denotes the organization which is formally 'responsible', while the term water manager implies the person or group of persons within the organization who 'implement' control actions.

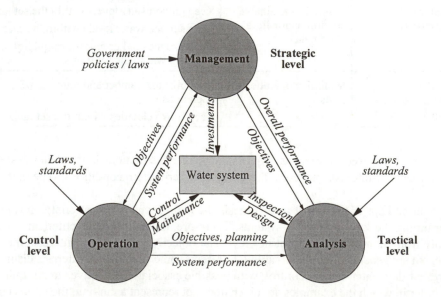

*Fig. 1.12. Decision-making levels and information flows within a water authority.*

Within the organization of the water authority, three organizational levels can be distinguished with respect to decision making: strategic, tactical and control (Fig. 1.12).

At the strategic level, interests are considered and objectives to be met in the water system are defined, on the basis of government policies and laws. The objectives, including their weighing, are usually fixed for the long term, e.g. in five-year plans. The management present at this level, is responsible for the overall performance of the water system.

The management verifies whether the objectives are met, on the basis of information supplied by the tactical and control levels. The strategic level is, moreover, responsible for investments in the water system and setting of priorities for structural measures.

Employees at the tactical level are responsible for planning, design and development of operational objectives. Within the limits set at the strategic level, the technical boundary conditions for operational control of the water system are set.

Infrastructural measures required are determined, incorporating the way in which control will be executed. In the designs, excess capacities are included, thus creating options for future extensions in the water system.

Operational objectives are developed on the basis of the objectives set at the strategic level that are valid for mid- to long terms. From the system-performance results, management information is assembled. Additionally, it is evaluated how well the objectives are met, on the basis of results obtained during the course of time.

The control level concerns the day-to-day operation of the water system. The employees at the control level are responsible for controlling the water system. They execute control actions, monitor water-system behavior and take corrective measures if undesirable situations occur. They are furthermore responsible for maintenance of the various control-system elements. Employees at the control level should be able to unambiguously execute the operational rules that are developed at the tactical level.

General knowledge of water-system behavior is essential in decision making at the tactical and control levels and it is important that the employees can predict and analyze the behavior of the system before operational actions are taken. The methods described in this thesis are specifically meant to support the work at these two levels.

## 1.5  Comparison with Similar Problems

Interest weighing in an operational situation, in which dynamic control is applied, requires optimization in selecting one or more of the possible control actions. Optimization methods have been applied in the process industry for a long time. In water management, large-scale applications are found all over the world. In the USA for instance, entire rivers and storage reservoirs have been controlled on the basis of the outcomes of optimization models for decades. In these projects, generally, various interests have to be satisfied, such as hydro-electric power, agriculture that is dependent on irrigation, fishery, etc. Usually, the objective is to maximize energy production by optimum water releases, while at the same time meeting the requirements of the other interests. This weighing problem has to be solved especially in periods of extreme hydrological conditions such as excessive precipitation or drought.

The operational problems involved in these large-scale systems resemble the problems encountered in the present study. A mix of interests has to be satisfied and too many or too few resources have to be distributed in a balanced way.

In addition to similarities, there are clear differences between the present subject and those studied in other countries. The most striking difference is the scale of application. In the USA in particular, large rivers and storage reservoirs are controlled, whereas this thesis deals with much smaller hydrological units such as storage basins, small rivers or creeks.

Furthermore, the time scale of the problems differs. Most applications of optimization for storage-reservoir control define the optimal releases for an entire year. The successive stages of the problems defined usually cover one month, reservoir water releases for each month being the outcome. Here, the period for which optimal control has to be determined, covers only a few hours to several days.

In general, hydropower generation is an interest of great importance in reservoir control and therefore great efforts are made to model that interest. Even the choice of optimization method seems to be influenced by this interest. In the Netherlands, hydropower generation is not an issue in regional water systems. However, the number of interests that has to be incorporated in control decisions in the Netherlands is large, while the exact definition of interests and associated requirements is often less clear and sometimes even vague.

## 1.6   Concluding Remarks

At present, a growing number of sometimes conflicting interests has to be taken into consideration in water management. Water systems developed during the history of land reclamation are basically not designed to incorporate the wide range of requirements and restrictions associated with the newly defined interests.

The excess capacities normally built into the current water systems has allowed postponement of extensions for some time. However, the capacity limits of many of these water systems are now reached and additional measures will be necessary to enable a water management, which meets the requirements of various interests.

Two fundamentally different measures can be considered: extending system capacities, possibly combined with separation of interests, or improving the performance of the existing water system by improved control.

Extending the present system capacity is expensive and so is separation of interests into hydrologically independent units. Separation is sometimes not even preferable because of historical reasons, or the fact that the value of one interest can enhance that of another.

An alternative for capacity increase and separation is improving the use of currently available resources in water systems by enhanced control. Such control incorporates the time-dependent interests defined and the water-system dynamics. This is called dynamic control. Various forms of dynamic control have shown good results in water management in several countries. In the Netherlands there is no experience in this field as yet.

The main objective of the present study is to determine universally applicable methods for of regional water-system control, including both rural and urban areas. The subject of improving the control of urban sewer subsystems has been studied extensively by others (e.g.: Einfalt et al., 1990; Hartong & Lobbrecht, 1992).

    This thesis presents a new approach, in which all subsystems of a water system are controlled in an integrated way. This is based on explicit weighing of interests, possibly crossing the borders of the responsibilities of different water authorities.

# 2    Decision Support and Control

## 2.1    Introduction

This chapter describes the elements of control systems that are relevant to the present study, in particular, systems that are designed to support the water manager in controlling his water system. These systems are called *Decision Support Systems* (DSS).

In general, the term DSS is used for a wide range of systems. It is used for simple decision trees and for computer programs to solve complex problems. In this thesis, a slight variation of the definition of Sprague and Carlson (1982) is applied:

'A Decision Support System is an interactive computer-based system that helps decision-makers to utilize data and models to solve complex problems.'

Several comments in regard to this definition should be made. Firstly, a DSS is interactive, i.e. there is a dialog between user and computer. The computer produces 'answers' to 'questions' asked by the user. Such questions can be in the form of: 'What if I take this action' or: 'What if this happens', with the computer presenting results on which decisions can be based.

Secondly, a DSS is used by decision-makers who, in this specific case, are those persons of a water authority responsible for water-system planning and operation.

Thirdly, the DSS itself is not a model but a computer-based system, which uses data and models, that are specifically intended to assist in the decision-making process.

Finally, a DSS is needed to solve complex problems. The complexity can follow from the water system that has to be controlled which, for instance, comprises many subsystem interactions, or the difficulty in satisfying the requirements of all interests.

In general, a DSS is needed if the water manager wants to obtain the best solution to his problem and a logical and an unambiguous choice from various solutions to that problem cannot be easily determined without a computer-based system.

A DSS can be used at different organizational levels (see Sec 1.4). In this thesis, only the control and tactical levels will be considered. At the control level a DSS can be used for real-time control, whereas at the tactical level it can be used for analysis purposes. In the latter case, a DSS can be used to support the work of planners, for instance, in the water-system designing process.

*Control systems* can be defined as systems that steer processes and correct discrepancies ,between actual and desired system states. The following general overview describes the main functional elements of a control system and its data and information flows.

For the general understanding, it is important to realize that the water manager has various sources of information, e.g.: observed data, automatically gathered monitoring data and system-load predictions.

The quantity of information required to obtain a balanced form of water-system control depends on the type of water system, its runoff characteristics and the required level of control. These sources of information can be used in models to obtain a complete overview of the water-system state and an estimate of the state for some time ahead.

## 2.2   Developments and Use

The first DSSs for water system-control were developed in the mid-seventies (Loucks, 1991). In the early stages, the approach focused on modeling and specific problem solving. These systems lacked the fancy interfaces which are common today. Data were entered into the systems by typing values. The output of such systems consisted of a lot of printouts, which could only be interpreted by the persons who had built the model.

At present, user-oriented, interactive systems are evolving, which enable a direct dialog between user and computer. User actions trigger computer actions and a process of questions and answers guides the user through these systems. Graphical output is commonly applied. Standardization of interfaces enables users who do not have any experience with a specific system to learn it rapidly.

As mentioned above, a DSS has to be used at various levels that exist in a water authority. Of all three organizational levels mentioned in Sec. 1.4, the DSS intended in this thesis is used at the tactical and control levels.

At the tactical level, a DSS should provide information which enables plans to be made and evaluated. The system might not be used frequently, but when it is used, the objective is to determine the long-term effects of certain measures. Furthermore, the DSS is used to assess the performance of control, to eventually develop new guidelines for the design and operation of water systems.

At the control level, the frequency of using a DSS depends on whether it is used for real-time control or not. If the DSS is used for real-time control, its operation is needed continuously whereas, if it is used off-line, its frequency of use is low and dependent on the analyses required. In general, the level of detail required in the operational situation will be high and the time span involved short, e.g. hours to several days. In addition, details of short-term and long-term historical events are of interest at this level, especially to determine future operational actions.

## 2.3    Levels of Control

A general overview of a water-control system is given in Fig. 2.1. The figure is a further specification of the modern control system as described in Sec. 1.3.2. Three levels can be distinguished in a control system:

*   decision-support level,
*   central-control level,
*   local-control level.

Control actions can be carried out from each of these levels. The person or device eventually taking the operational decision, depends on the current status of the water system and the risks which are involved in the operation. Note that the decision-support level and the central-control level are located centrally, in general at the same location. The local-control level is in the water system.

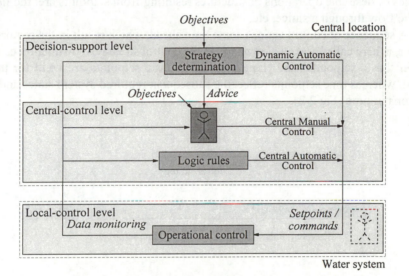

Fig. 2.1. Principal levels in water-system control.

### 2.3.1  Decision-Support Level

At the decision-support level, decisions are prepared, analyzed and evaluated. The objectives for water-management form the basis in decision making. These objectives should be formulated in such a way that employees at the tactical and control levels can base their decisions on it. An example for an area that incorporates nature, agricultural and recreational interests could be: maintain conditions in which natural values can develop best as well as agricultural production; prevent flooding of agricultural land as well as poor water quality in recreational waters.

The process of formulating objectives, implies a form of interest weighing. Since it cannot be known in advance whether the objectives stipulated will result in the water management intended, evaluation is necessary before the objectives can be applied in practice.

The decision-support level can be used in various ways: planning, analysis and real-time control. An example of the use for planning purposes is the design of a water system. An example of the use in analysis, is determining the performance of a controlled system on the basis of the objectives set. In case of real-time control, the desired system state is determined at the decision-support level on the basis of actual data.

The desired system state is generally obtained by a combination of operations of regulating structures. The desirable water-system states and regulating-structure operations are here formulated as *setpoints*. Examples of desirable system states formulated as setpoints are: a particular surface-water level; a particular groundwater level; a chloride concentration limit. Examples of desirable operations of structures resulting from setpoints are: the flow over a weir; the flow through a sluice, etc.

A sequence of control actions some time ahead, which result in the required system behavior, is here called a *control strategy*. In the control strategy, a distinction can be made between 'basic setpoints' and 'time setpoints'. Basic setpoints are valid for the current moment, whereas time setpoints reflect the subsequent required system states and structure operations in time (Fig 2.2).

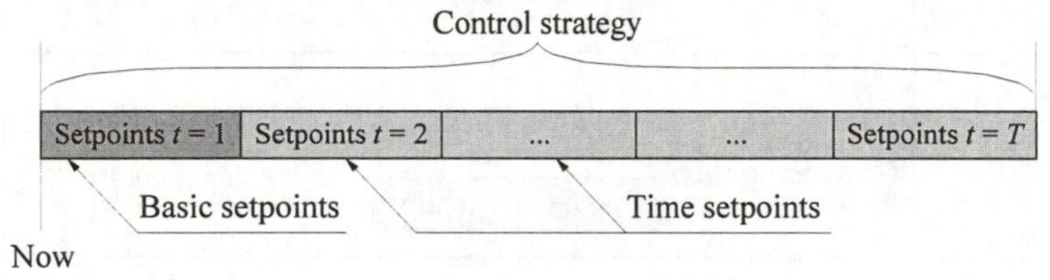

*Fig. 2.2. The control strategy as a series of basic setpoints and time setpoints.*

Once a control strategy is determined, it can be applied directly from the central location. Such a control strategy can be in the form of an advice to operators. When their operational actions are based on that control strategy, the control mode is called *dynamic manual control*.

The control actions required, can also be sent to the local-control level directly, by transmitting basic setpoints, which are implemented automatically. This is called *dynamic automatic control*.

## 2.3.2  Central-Control Level

At the central-control level, operators of a water system take action to maintain desired water-system states. Various forms of central control exist, ranging from remote manual control to automatic control by means of computers.

Two forms of central control can be distinguished (Fig. 2.1): manual control by operators, here called *central manual control* and automatic control based on logic rules, called *central automatic control*. In manual control, the required setpoints or control actions are determined by the operators in person. Central automatic control is based on logic control rules. By means of these rules the required control strategies are determined automatically for the various circumstances accounted for.

Both manual and automatic central control are forms of *remote control*. This means that the operating commands are issued at another location than the regulating structures which have to be controlled. In general, the central functions do not necessarily have to be located at a single place, but they can be distributed over several locations. For simplicity, the central location will be considered here to also represent these other possible situations.

## 2.3.3  Local-Control Level

At the local-control level, the required control actions are implemented and monitored. If the value of a water-system variable differs from the current setpoint, local controllers take action to maintain the setpoint. These controllers will be considered automatic in this thesis, while the setpoints are transferred to them electronically.

Monitoring both controlled and uncontrolled variables in the water system is necessary to evaluate whether the implemented control strategy complies with the objectives set. In general, it will not be possible to monitor all the variables required to achieve a complete picture of the water-system state. As will be explained later, some missing variables can be estimated artificially.

In controlling large water systems a further hierarchy can sometimes be observed which distinguishes more levels than the ones discussed above. In those cases the functionality of the central level can be replicated at the local level (the dotted operator in Fig. 2.1)

## 2.4  Control-System Functionality

In this section, the main functional elements of the three-level control system are discussed in more detail. Figure 2.3 presents the major elements and data flows in a control system. In addition to internal data flows, data from outside the control system are generally necessary.

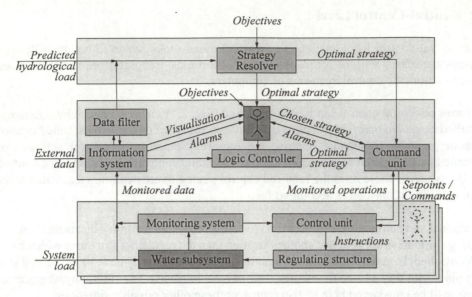

*Fig. 2.3. Major elements of a real-time control system and data flows.*

Table 2.1 shows the naming conventions for control, used throughout this thesis. A major distinction is made between carrying out operations automatically and manually. Automatic operations are executed by computers and electronic devices, whereas manual operations are implemented by operators of a water system.

The way in which the operations are implemented, is also reflected in the control names, e.g. *dynamic automatic control* is implemented automatically.

In dynamic control the dynamics of the water system and the mutual weighing of interests in the water system are incorporated in the control strategy. Since dynamic control uses information from several subsystems, commands are generally issued from a central location. In dynamic control, the control strategy is determined by a DSS and available to the operators as an advice or, optionally, applied directly by control units. Dynamic control can also be applied manually. In that case it is called *dynamic manual control*. In principle, dynamic manual control commands can also be applied locally. Depending on the number of regulating structures included in dynamic control, this can be a very laborious task.

In automatic operation, both the central and local levels implement control actions on the basis of logic rules, the relevant control modes are *central automatic control* and *local automatic control*.

In manual operation, heuristics can be involved in determining the best control strategy. To apply this successfully, the operators should have the required knowledge about the water-system behavior that will result from the control actions they take. Manual control commands can be given both at the central and local levels: *central manual control* and *local manual control*.

Table 2.1. Naming conventions in water-system control.

| Operation | Control level | Strategy determination | Control mode |
|-----------|---------------|------------------------|--------------|
| Automatic | Central | DSS | Dynamic Automatic Control |
| | | Logic | Central Automatic Control |
| | Local | Logic | Local Automatic Control |
| Manual | Central | DSS / Heuristic | Dynamic Manual Control |
| | | Heuristic | Central Manual Control |
| | Local | Heuristic | Local Manual Control |

### 2.4.1 Operational Conditions

Several operational conditions can be distinguished, irrespective of the control mode used to manage the water system:
*   standard operation,
*   anticipated system failure,
*   system maintenance,
*   emergency operation.

During standard operation, the control system is fully operational. Predefined operating conditions are available, each guaranteeing safe conditions of the water system and the control system. The rules that have to be followed, have to be described in an operations manual (Schilling, 1991).

In the situation of anticipated system failure, the control system is not fully operational due to accidental or unexpected events, which, however, can be explained, e.g. breakdown of a motor unit of a pump, communication failure, etc. Anticipated failure conditions and the actions to be taken by operators should be described in the same operations manual as standard operation.

Parts of the system that are closed for maintenance cannot be controlled. Maintenance should preferably be carried out during periods when low system loads are expected.

Emergency operation occurs when the system is not fully operational but the corresponding situation is not covered by regulations. Such failure situations are not included in the procedures for anticipated failure. These situations generally require manual control of the water system.

The first three operational situations are included and further discussed in this thesis in the descriptions of design and operation of water systems. In principle, for these operational situations, the DSS can be used to support operators in determining the best control strategies for the water system.

### 2.4.2 Strategy Determination

The main focus of the present study is on strategy determination on the basis of a DSS and on the basis of heuristics. Strategy determination on the basis of logic rules is not discussed in detail.

At the decision-support level, suitable control strategies are determined within the limits and objectives of the current policy and the given water-system state. The best strategy possible, is called the 'optimal strategy'. The control strategy can be obtained in two ways:

- by trial and error;
- in a single run.

Figure 2.4 shows the principle of both methods of strategy determination that can be applied at the control and tactical levels in a water authority (Sec. 1.4). At the control level, its use is primarily for real-time control. In addition, it can be used for training and analysis purposes. At the tactical level, its use is primarily for analysis purposes, to determine the water-system performance for longer periods and for planning purposes.

Determining a control strategy by trial and error, requires a calibrated simulation model, called a *Simulator*, which reflects the behavior of the real-world water system. When the single-run method is used, a system called the *Strategy Resolver* needs to be run once (see also Fig. 2.3) This system determines the optimal control strategy on the basis of objectives formulated.

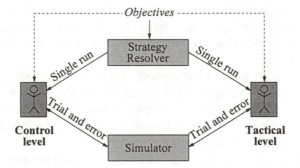

*Fig. 2.4. Strategy determination at the control level and the tactical level.*

The control strategy describes a set of control actions some time ahead. The control strategy depends on the response time of water system and the control actions. In water systems with rapid runoff characteristics a prediction of the hydrological load is generally necessary. In water systems with slow runoff characteristics a prediction is not always necessary.

The period for which the control strategy has to be determined is called the *control horizon*. The control horizon can be seen as a time window that has a fixed length, which shifts in time. The required length of the control horizon is to a large extent determined by the runoff characteristics of the water system. This will be discussed further in Sec. 5.3. For the moment, it is assumed that the length of the control horizon is known and that, if necessary, a prediction of the system load is available.

The steps of strategy determination processes according to the trial-and-error method and the single-run method are shown in Fig. 2.5.

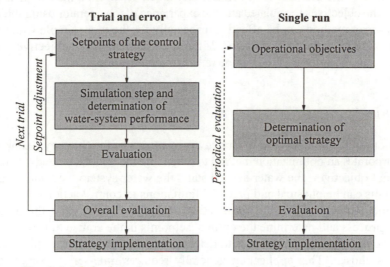

Fig. 2.5. Methods to determine a control strategy: by trial and error and in a single run.

### By Trial and Error

Using the trial-and-error method, first a set of initial setpoints to control the water system has to be selected for a particular time span, e.g. one hour. Using the selected setpoints and the predicted water-system load, the water-system behavior is simulated for one time step. After this step, the performance of the water system is evaluated against the objectives. If there is a risk that the simulated water-system state deviates from the desired one, some, or maybe all the setpoints are adjusted before simulating the next time step. The adjustments are determined on the basis of knowledge and experience. Again, the water-system behavior is evaluated. This process continues until the entire control horizon has been simulated.

The setpoints chosen for the various regulating structures during the control horizon, reflect the attempted control strategy. This strategy is subsequently matched with the objectives, considering the entire control horizon. It should be mentioned that this overall evaluation is similar to the evaluation that is done after each simulation step, but it incorporates all water-system states during the control horizon. The evaluation result of separate moments in time on the one hand, and the evaluation result of all moments in time together on the other, may differ with respect to the overall water-system performance. For instance, in a particular time step, a poor performance may have been chosen, in favor of a next step with a very good performance.

In general, the procedure will be followed using several simulation steps at a time. After the overall evaluation, extra trials may be necessary to assess alternative control strategies and determine which one is best.

The effectiveness of the trial-and-error method very much depends on the complexity of the water system, the objectives formulated and the experience of the operator using the method. The trial-and-error method can be effective, especially for the more simple water systems where setpoints rarely have to be adjusted. If the water system or the objectives are more complex, the trial-and-error method can be rather time-consuming.

### In a Single Run

Another option to determine a control strategy is by means of the Strategy Resolver. This resolver incorporates an optimization method to determine the optimal strategy, considering the operational objectives, the water-system state, the water-system load and constraints. Such constraints can be physical and practical limitations to control actions.

This method requires careful formulation of the objectives, since these guide the optimization process and determine the optimal setpoints in the course of time.

The use of optimization methods to determine control strategies, implies weighing the objectives chosen. This application generally requires time-series calculations and analyses to check the effectiveness of the optimization process in the long run.

Optimization is part of the Strategy Resolver that incorporates a model of the water system. Its operation can be compared with an evaluation of all possible setpoints for regulating structures during a particular control horizon. However, the resolver computes a limited number of alternatives, omitting alternatives that result in a poor overall performance.

The single-run method is fast in comparison with the trial-and-error method. This is advantageous in real-time control. The single-run method should, however, be evaluated periodically, to determine whether the method still reflects the current policy.

### Comparison

The trial-and-error method only permits manual control of a water system. Because of the time involved in determining a good control strategy, the trial-and-error method in a real-time situation may not always be appropriate.

The computer-based single-run method enables both manual and automatic (see Fig. 2.1) implementation of the control strategy. In case of manual control, the operators themselves determine whether and how to implement the control strategy. This makes it possible to incorporate specific situations or temporary changes in the required system states. However, the single-run method also has some drawbacks. These will be discussed in Sec. 3.2.1.

### 2.4.3 Central Control Tools

At the central-control level, operators can have various tools available to facilitate their operations (see also Fig. 2.3):
• an information system;
• a logic controller;
• an alarm system.

Data are processed into information by an 'information system'. Processed actual and historical data can be made visible via the information system. By means of interactive dialogs, the operator can retrieve detailed information about the ongoing processes in the water system.

In general, not all required data will be available from on-line monitoring. Some data can be measured manually at relatively long intervals, e.g. water quality data can be obtained from periodical measurements and laboratory analyses. Such data have to be supplied from outside.

A method to overcome a lack of continuous monitoring data is to use models, which simulate the water-system behavior and attribute values to missing variables. Such models should be calibrated regularly, or even on line, to keep producing accurate results. An option for automatic calibration is the use of a numerical filter that makes sure the model keeps resembling actual monitoring data. This type of data processing, using both measurements and models is known as 'data assimilation'.

If balanced water-system control can be obtained by logic control rules, they can be applied in a *logic controller*. The control rules can result in flexible strategies, fully determined by the actual water-system state, or in a fixed set of 'control recipes', which cover all the control actions necessary to obtain the required water-system behavior. If the logic controller is used, the functionality of the control system generally allows automatic and manual selection of control strategies.

Use of the logic controller always requires matching the situations distinguished in the controller with the actual system state. This form of determining a control strategy is also know as the 'verification method' (Sec. 3.2.1).

An operator is warned if undesirable situations occur in the water system, e.g. too low or high water levels, or too high pollutant concentrations. The alarm in that case is issued via the information system, where such data are available.

One of the tasks of the central command unit is to monitor regulating structures, operated by local control units. When one or more of the control units or regulating structures is malfunctioning, a local alarm is issued and sent to the command unit. Subsequently, the responsible operator is warned.

### 2.4.4 Command and Control Units

Operations of regulating structures are activated through instructions. These instructions are given by command and control units. The control system of Fig. 2.3 shows a central command unit and a local control unit. In general, there can be several local control units, which regulate structures or entire processes. A local control unit can function as a central command unit for other local control units. Entire hierarchical networks of control units are built in this way. Only two levels are discussed here: one central level and one local level.

The central command unit should be seen as an agent which controls processes via local control units and checks their operation. The centrally determined control strategy is transmitted in the form of setpoints to the local control units. Upon receiving these data these local control units take care of the local processes. The local units are monitored continuously. If an undesirable situation occurs or one of the local systems fails to operate properly, an alarm message is issued and sent to the operator.

The local control unit gives instructions to regulating structures, e.g.: to the motor-driven device that controls the crest of a weir; to an engine of a pumping station to switch on or off

Contrary to the decision-support and central-control levels that are often located at the same place, local systems are generally located at various places all over the water system. Data communication between central and local systems and between local systems, takes place via a *communication network*. There are various options for this type of communication. Of importance in the current discussion is that such a network may break down, preventing setpoints from being sent to the local level.

To deal with such breakdowns, special strategies should be developed and implemented by default setpoints for all local control systems. The default setpoints should guarantee a permanent safe control and should be implemented automatically. This is also known as 'fail-safe operation'.

A modern control system generally includes an on-line monitoring network that automatically gathers data from the water system. In general, local control units are equipped with monitoring devices, e.g. water-level measurement equipment. However, additional monitoring locations may be necessary to obtain the required picture at the central level.

The continuous monitoring of variables in a water system enables determination of the actual water-system state. These variables also give an indication for the expected runoff or other processes.

In addition to monitoring variables within the water system, it may be an option to monitor the hydrological load on the water system. Sometimes anticipation of runoff or a lack of runoff, enhances the water-system performance. In that case, monitoring of the hydrological load and simulation of the water-system behavior, yields a prediction of water-system variables. This enables anticipative control actions, which is of special interest to those water systems that have rapid runoff characteristics.

## 2.5 Concluding Remarks

A Decision Support System (DSS) can be used in water management as a tool to enhance the work of the water manager at several organizational levels. The present study focuses on the tactical and control levels. At the tactical level, plans are made for the mid to long term. For those purposes the analysis options of the DSS can be used. At the control level, the DSS can be used as part of a real-time control system, to improve day-to-day water-system operation.

The strategy for controlling the water system can be obtained by trial and error using a Simulator, or in one run, using the Strategy Resolver. The Strategy Resolver is especially worthwhile in water systems which are complex to control and in general for the application of real-time control.

If used for real-time control, the Strategy Resolver advises operators who implement control actions themselves or controls the water system directly. If the required control of the water system allows so, the control actions of the Strategy Resolver can be abstracted into logic rules which can be implemented in a logic controller and implemented automatically.

In practice, operators may have several other tools to support them in their work. If such tools do not function on-line, operators will probably not use them when extreme system loads require fast action. In such cases, the water system will be operated on the basis of operators' experience and knowledge.

Uninterrupted functioning of the entire control system can never be guaranteed. A situation can always occur in which manual control is necessary, even for entirely automated control systems. In such cases, the experience and system knowledge of operators is of decisive importance to achieve the correct control.

To enhance the water-system knowledge of operators, the Strategy Resolver can be used as an off-line training tool. Such a tool can also enhance their work in assessing automatically implemented control actions.

The operation of a water system is generally determined by the risk involved in control. Under normal conditions, the Strategy Resolver or a logic controller can determine all control actions, without operator interference. Under extreme conditions, however, the situation in the water system may become so critical that the control actions at least have to be supervised by operators. In such critical cases, an operator can decide to use the strategy advised by the Strategy Resolver, the strategy of the logic controller, or if necessary, determine a control strategy himself.

# 3 Problem Formulation and Solving Methods

## 3.1 Introduction

Dutch water systems have traditionally been controlled by structures that have fixed setpoints or control ranges for the summer and winter season. Formerly, these fixed setpoints were obtained by weighing the interests present in a water system on the basis of the output of each interest over an entire season or even several years. In the past, when most interests in water systems had an agricultural character, this weighing was a question of balancing the economic output of an area.

As described in Sec. 1.3, this form of water-system control does not meet the present-day objectives nor the future ones probably. In the present study, various options to improve operational control are presented as an alternative to building extra infrastructural capacity in the current water systems.

The new operational control options should include the operational flexibility that exists in the current water systems, specifically the permissible fluctuations in groundwater levels, surface-water levels and water quality.

The objective of this study is to achieve a flexible water management that incorporates the time-varying requirements of the different interests that are present in a water system, the predicted system load and the dynamic behavior of the water system.

Meeting interest requirements can often be expressed in terms of physical water-system variables. In the following, specific examples of such expressions are given for polder as well as for hilly types of areas.

Agricultural requirements can for instance be expressed in groundwater level, a soil-moisture-content range, a minimum surface-water level and/or a maximum salt concentration of irrigation water. The groundwater level and the soil moisture content determine the accessibility of the land for machinery, the crop-growing conditions and in drought-sensitive areas, the need for sprinkler irrigation. Irrigation water can be abstracted from the groundwater and from the surface-water subsystems. The latter requires a minimum surface-water level. Furthermore, the salt concentration of this water should not exceed certain limits, to prevent reductions in crop yields.

Recreational requirements may be expressed in required surface-water levels, surface-water quality and limits to surface-water velocity. Maintaining surface-water levels within certain limits can be of importance to recreational navigation (e.g. heights of yacht jetties and bridges). Surface water which is used for swimming, should meet specific water-quality standards. Surface-water velocities have to be low enough to allow swimming and navigation.

Aquatic ecological requirements may be an undisturbed and stable surface water quality and maintaining a minimum surface water level. Especially during spring, water quality is of extreme importance, since flora and fauna start to spring up. Flushing surface water should be prevented as much as possible, in principle during the entire year. A problem is that, especially in drought-sensitive areas, canals may run dry without external water supply. This may necessitate water inlet into such canals. Because of these problems, weighing may be required within the aquatic ecological interest.

Nature requirements may not be very well-defined. Some people define nature as the situation that exists or develops when areas are left untouched by humans. In the highly developed Netherlands there is hardly a square meter left that satisfies this requirement and the question can be asked whether the nature that would develop, if certain places in our current environment were left alone, is really appreciated so much. In general, the opinion is that some kind of human interference will continue to be necessary in nature reserves. Nature requirements expressed in water-control variables, are limited to permissible ranges for groundwater and surface water levels and restrictions with respect to water quality. Depending on the location and the type of nature that should develop or sustain, the requirements vary from very strict to not really limited.

In municipal areas the main concerns are to maintain the water-quality standard of the surface water and to prevent flooding. Maintaining a good water quality in the surface water subsystem of a municipal area may require regular flushing and continuous water movement to keep oxygen levels above the minimum limits to prevent stench. Moreover, during and after sewer overflows the water quality may become rather poor. This requires additional flushing. Furthermore minimum and maximum groundwater levels may be required. Maxima are generally determined by the depth of cellars that should not flood. Minima may be determined by timber piles in old cities that may decay under aerobic conditions.

Navigation for transportation purposes, requires minimum and maximum surface water levels. Too low a water level may cause a ship to run aground. Too high a water level may cause a ship to hit a bridge.

The assumption in this study is that the requirements of each interest can be expressed in physical water variables of rural and urban subsystems which can be directly or indirectly controlled, e.g.:
- surface-water levels,
- groundwater levels,
- sewer filling,
- water quality level.

A method is presented to solve the operational control problem of the water manager, meeting the requirements of all interests present in a water system as well as possible. The operational control problem in this context is solved by determining the best control actions for the moment and for some time ahead, on the basis of given historical, present and predicted system loads. A set of control actions some time ahead has been introduced as the control strategy (Sec. 2.3.1) and the period ahead as the control horizon (Sec. 2.4.2).

The optimization method was chosen to determine the best control strategy at any moment. Other methods, such as the trial-and-error method (Sec. 2.4.2), which include iterative calculations to determine the desired control strategy, are not examined.

The main reason for choosing optimization was that it makes it possible to determine control strategies in a single run. Fast determination of control strategies is of decisive importance in real-time control of water systems, where rapid operational decisions are essential.

The method developed here uses the flexibility available in the current water systems. Elements of fixed targets or control scenarios can be included. Such elements force the optimization process to search in a narrow solution space, however.

A weighing mechanism is introduced to determine the relative importance of interests. This mechanism includes the time variability of interest requirements during the year.

Several methods for finding the best control strategy are discussed in this chapter. The methodology developed for solving the control problem consists of: an overall objective for the water system; relationships between water-system variables and limits to variables defined. Methods for solving such problems are called 'methods for multi-variable search' in literature. This term stresses that the optimal values of variables have to be found, generally within a limited space.

Here, the method for multi-variable search will be called the *optimization method*. The formulation of the objective of control, the water-system relationships and the limits to water-system variables together form the *optimization problem*.

Optimization problems generally have many possible solutions, of which one or more can be optimal. An ideal solution, which cannot be improved upon, is called a *global optimum* in literature. It is not always guaranteed that the global optimum will be found during the search. An optimization method may be trapped during its search process in a *local optimum* (Fig. 3.1). In such a case the method may indicate that an optimum is found, which is, however, not the global optimum. In optimization it is very important to be certain that the global optimum is determined and not a local optimum, since the latter may be far from optimal.

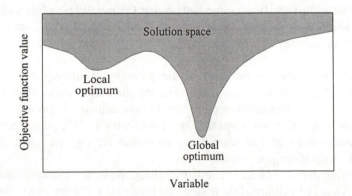

*Fig. 3.1. Local and global optima in a process of minimization.*

## 3.2    Solving the Control Problem

### 3.2.1  Optimization Methods Applied

Several methods are applied to solve operational control problems. Basically, all are optimization methods. In general, the aim of applying these methods is to find the best control strategy. The main methods discussed here are:

- heuristic rules,
- verification,
- neural networks,
- genetic algorithms,
- mathematical optimization.

Heuristic rules are based on experience. Heuristic rules describe how the water system should be controlled to achieve a high water-system performance. This method has a wide range of applications. It varies from simple operational rules on paper, to rules implemented by computerized knowledge-based systems.

Simple heuristic rules are successfully applied in real life. They limit the variety of possible operational decisions, which is generally considered an advantage. Knowledge-based systems have been proven to be capable of reproducing satisfactory results (Babovic, 1991). Heuristic-rule-based systems have to be updated as soon as one or more elements of the water system change. They have the advantage, however, that the reasoning mechanism used, can be made clear to the end user.

Verification is based on comparing system loads and water-system states in known situations with new situations. Several verification methods exist. Most popular is the use of a 'decision tree', using if-then-else structures, which can be set up to deal with a variety of circumstances.

Another form of verification is the matching method in which possible and known water-system states are listed and the state which best matches the current state is identified. The control strategy corresponding to the best-matching state is chosen. A practical disadvantage of this method is that it is difficult to match actual events with known events.

The advantage of verification methods is that the possible control strategies can be determined in advance, using deterministic models. Verification methods are generally fast, which makes them very suitable for real-time control.

Neural networks replicate the behavior of the brain by emulating the operations and connectivity of biological neurons. Such networks are often regarded as black box methods. In a neural network, a series of connecting weights are adjusted in order to fit a series of inputs to another series of known outputs. When the training set of a neural network is large enough, the system is capable of reproducing an output for a given input, if this input is included in the original range of validity.

The training period of a neural network is long, while its response time is generally very short. The range of validity of the training set is of major importance to the application of neural networks. Situations for which the network has not been trained, e.g. certain water-system loads and states, should be prevented. In case the water-system arrangement changes, a new training set is required and the network has to be trained again.

Genetic algorithms are random-search methods that simulate the mechanics of natural selection and genetics. A genetic algorithm selects a number of promising control actions, which together form a control strategy, applies them via a simulation model and evaluates the results in order to generate a new control strategy. Each control strategy is called an 'individual structure'. Each new individual structure is determined by combining parts of the most successful ones, called 'cross-over'.

The algorithm maintains a population of individuals that evolves according to certain rules of selection. Each individual is assigned a performance rating, called 'fitness'. The fittest of each population are kept, the others discarded. After a generally large number of iterations the method produces convergence to an optimum. There is no guarantee that this is the global optimum though.

Genetic algorithms are studied widely these days, also for solving water-control problems (e.g. Esat & Hall, 1994). No real-time control implementations have been reported yet.

Mathematical optimization is used if the minimum or maximum of a function has to be determined. This function is a mathematical description of the operational objectives. Application of this method requires the objectives to be aggregated into mathematical descriptions in terms of cost of either achieving or violating the given requirements. One of the shortcomings of this method is that it may require a high computational load. Depending on the specific form of the method, it can be guaranteed that the global optimum is found.

All methods have their own capabilities and limitations. A detailed comparison and evaluation has been made by Martín García (1996).

Heuristic rules are easy to implement, but they are only valid for one specific water system and do not ensure optimal control actions. An advantage, this method shares with the verification method, is that the way in which a decision is made can be traced.

The principle of the verification method is simple, but the number of possibilities that can be expected, can become very large. One of its applications can be controlling a water system by means of logic rules, as introduced in Sec. 2.4.3.

Neural networks are becoming popular. Advantages of neural networks are that it is not necessary to know the problem structure and that the method is very fast, once the network has been trained. The method requires a model of the water system or detailed monitoring data and associated control actions. The fact that a new training procedure is required each time an element of the water system changes, is a major disadvantage. Another disadvantage, this method shares with genetic algorithms and to a lesser extent with mathematical optimization, is that the reasoning behind decisions cannot be traced easily.

The application of genetic algorithms seems to be very promising, although the process of convergence is generally very slow.

Mathematical optimization finally, enables determination of optimal control strategies in a more deterministic way, but can be computationally demanding and requires building a mathematical model that closely resembles the real water system.

### 3.2.2 Mathematical Optimization

Without further examining and evaluating the other methods, in this thesis, mathematical optimization was opted for. This method has advantages, some of which are missing in the other methods, which are of decisive importance for water-system control. The major advantages of mathematical optimization are:
- it is universally applicable;
- it includes deterministic modeling;
- it is robust;
- it enables relatively easy model building;
- it is already widely applied.

With respect to universality it can be mentioned that mathematical optimization can be used to solve a wide range of problems. Once the structure of a water system is known, a mathematical model can be built, using standard modeling rules for polder and hilly areas.

Mathematical optimization is deterministic, which makes the underlying processes of finding the optimum understandable. A mathematical optimization problem can be analyzed and the model behavior can be explained without having to solve the entire mathematical problem. Unexpected model behavior can be traced by analyzing the way in which processes are modeled. This option can enhance the credibility for end users.

Exact reproduction of results which were obtained previously is of great importance in water-system modeling for real-time control. The robustness of mathematical methods is the result of the exact formulation of the problem by means of mathematical relationships.

Mathematical optimization problems can be easily built and changed by means of special software. Such software tools enable efficient modeling, using predefined structures for the various elements of a water system. This feature makes mathematical models easily extendable as well.

As will become clear in the next sections, mathematical optimization is applied successfully for various problems similar to the one defined in the present study. This was considered an important criterion, since it comprises combining and enhancing state-of-the-art methods, which enables improvement of performance and robustness.

However, a drawback of the method is that operators at the control level of the water authority, have to have confidence in the outcome of optimization, because the reasoning of the method is not always entirely clear in the operational setting. This problem can be reduced by intensive off-line use of the method by operators (e.g. in training), to gain confidence in the outcomes, especially under extreme conditions in the water system.

## 3.3   The Optimization Problem

In describing a physical water system, a distinction can be made between *state variables* and *control variables*. State variables describe the state of the water system, e.g. water levels and pollutant concentrations. Control variables are the variables that are used to control the water system e.g., transport of water and pollutants through pumping stations. The general mathematical form of the optimization problem to be solved is given by:

$$\text{Minimize:} \quad Z(\bar{x}, \bar{u}),$$

$$\text{subject to:} \quad g_i(\bar{x}, \bar{u}) \leq 0, \qquad i \in 1, .., l;$$

$$x_{l,j} \leq x_j \leq x_{u,j}, \qquad j \in 1, .., m; \qquad (3.1)$$

$$u_{l,k} \leq u_k \leq u_{u,k}, \qquad k \in 1, .., n;$$

in which:

$Z(\bar{x}, \bar{u})$ : objective function;
$g_i(\bar{x}, \bar{u})$ : constraints;
$\bar{x}$ : vector of state variables;
$\bar{u}$ : vector of control variables;
$x_{l,j}, x_{u,j}$ : lower and upper limits of state variables $x_j$;
$u_{l,k}, u_{u,k}$ : lower and upper limits of control variables $u_k$.

The objective function of $Z$ in Eq. 3.1 represents the damage to the interests in the water-system. Two types of variables are distinguished: the water-system state and control variables, given by vectors $\bar{x}$ and $\bar{u}$. The number of constraints to the problem is $l$, the number of state variables $m$ and the number of control variables $n$.

The optimization process consists of minimizing the objective function subject to the constraints. These constraints $g_i(\bar{x}, \bar{u})$ describe the physical relationships in the system, e.g.: continuity for the surface-water subsystem (mass balance). The limits formulated, represent most upper and lower values of variables, e.g.: the minimum water level in a surface water subsystem; the maximum capacity of a pump.

In the problem defined, all water-system variables that play a role in runoff and control processes are included in Eq. 3.1 for an entire control horizon. By solving the optimization problem, the state and control variables are determined for each time step. The value of each control variable represents the optimum control situation for the corresponding regulating structure during a time step. The vector of all control variables represents the control strategy. Each state variable represents the optimal state of a water-system variable at the end of a time step. The vector of all state variables represents the entire water-system state during the control horizon.

## 3.4 Mathematical Optimization Methods

### 3.4.1 Types of Methods

In this section various mathematical optimization methods that are applied in practice are discussed. The objective is to give an overview of the most important aspects that have influenced the choice to use a specific method in the present research.

A mathematical model should be formulated in such a way that it can be solved by means of a mathematical programming algorithm. Such a mathematical model will be referred as *optimization problem* and its variables will be called *optimization variables*. The implementation of such an algorithm in computer code is called a *solver*.

The purpose of a solver is to find the optimal solution of the problem amongst all possible solutions. In general, the search space is limited by the constraints included in the problem. The more efficient the solver, the fewer evaluations of the search space have to be made to determine the optimal solution. If the optimum value of all problem variables is found, the search process is finished.

The search space contains valid solutions to the problem, one or more of which are optimal. However, in some circumstances it may be impossible to find the optimal solution. This situation can occur if the constraints set for the problem are so strict that no solution can be found, the problem originally formulated is then called *infeasible*.

During the past few decades, several optimization methods have been used to determine control strategies for water-resources systems. Yeh (1985) presents a state-of-the-art review of some of these methods, including simulation (trial and error) and mathematical-optimization (in a single run) approaches. Wurbs (1993) presents a more recent review. This section presents the results of a survey of widely applied mathematical optimization methods; simulation approaches are excluded from the discussions.

The main four mathematical optimization methods that are applied widely, are:
- Network Programming (NP), using a linear network-flow model;
- Linear Programming (LP), using a linear model;
- Dynamic Programming (DP), using a recursive model;
- Nonlinear Programming (NLP), using a nonlinear model.

An introduction to these methods can be found in Hillier & Lieberman (1990) and some applications in the field of water resources analysis in Loucks et al. (1981). Studies of water-resources literature incorporating mathematical optimization show that most applications are found in the USA (Vredenberg, 1996; Smakman, 1993).

In addition to the above methods, Successive Linear Programming (SLP) is distinguished as a separate method here. This method is generally classified as NLP in literature.

The choice of optimization method depends on two factors (Yeh, 1985):
1. the characteristics of the system under consideration;
2. the required modeling accuracy of the objectives.

Firstly, it should be possible to model the characteristics of the system under consideration with sufficient accuracy. Since a model is a simplified description of reality, important information about reality may be lost during modeling. When too much information is missing, the solution to the optimization problem is not accurate enough or even invalid.

Secondly, it should be possible to model the objectives with sufficient accuracy. Inaccuracy in modeling the objectives also results in an inaccurate solution.

Future requirements for model development and additional system features, are incorporated with relative ease (Wurbs, 1993). In the future, a water system may be extended with, e.g. pumping stations, weirs and inlets. Furthermore, objectives may change and the number of objectives may expand. These changes then have to be incorporated with sufficient accuracy in the optimization-problem description.

Limitations such as computer-memory size and calculation time, can make it difficult or even impossible to use methods that require extensive and very accurate models. If a water system consists of a large number of reservoirs, the computer-memory requirement and calculation times can become too large to be handled by a Personal Computer. This problem will soon be a thing of the past, because of the rapid developments in computer power and memory availability. However, the requirements of modelers also increase.

In real-time control of water systems it is necessary to calculate a new solution to the control problem regularly, incorporating the latest field measurements and hydrological-load predictions. This practice requires calculation times which are much shorter than the real time. If the control problem is not solved much faster than real time, the solutions may be outdated by the time they are available.

There are three options to incorporate, possibly conflicting, objectives in the optimization problem. In the first option, all objectives are incorporated in the objective function $Z$ with related weights to include their relative importance. This can be done, by using the principle of *Nonpreemptive Goal Programming*. For each objective, a specific numerical goal is set and an objective function is defined. The optimal solution is found by minimizing the weighted sum of deviations of objective function values from their respective goals.

In the second option the objectives are divided into different-priority objectives depending on their relative importance e.g., first-priority objectives, second-priority objectives, etc. For each objective, a numerical goal is set and an objective function is defined. The initial focus during optimization is to approach the goals corresponding to the first-priority objectives as closely as possible. If the optimal solution associated with the first-priority objectives is not unique, the second-priority objectives can be taken into account, while maintaining optimality for the first-priority objectives, etc., until a unique solution is found. This approach is called 'Preemptive Goal Programming'.

The third option can be used if one objective is essential, this is then used to define the objective function, while the other, less important objectives are included via constraints. Such methods are called 'constraint methods'. The constraint method gives a lower limit for a maximization objective and an upper limit for a minimization objective, so that a certain level of optimality is guaranteed for each less important objective. Care has to be taken when setting these limits since the problem might become infeasible if the limits are too strict.

In the following subsections the optimization methods Network Programming, Linear Programming, Successive Linear Programming, Dynamic Programming and Nonlinear Programming will be discussed. The following order has been chosen: from high modeling restrictions and easy to understand to high modeling freedom and complex.

For each method, the modeling characteristics are discussed first. Subsequently a few applications of the method in operational water control found in literature are described, giving the most important reasons for choosing the method in practice. Finally, some advantages and disadvantages of each method are discussed. This section ends with a summary of some key features of the various methods.

In the discussion, the term *dimensionality* is frequently used. It is used to describe the requirement of computer resources, e.g. memory and processing time for model evaluations of a solver. Dimensionality depends on the number of variables $(m+n)$ and constraints $(l)$ (Eq. 3.1) and on the level of nonlinearity, nonconvexity and discontinuity in the optimization problem. Each solver uses different data structures to represent the optimization problem, which results in different computer memory requirements, dimensionality therefore also depends on the solver and the algorithm applied.

In general, dimensionality can be defined as a measure that relates computer resources of a solver to the size of an optimization problem and its level of nonlinearity, nonconvexity and discontinuity, in which the size of the problem is measured by the number of optimization variables and the number of constraints.

In the current problem formulation the number of variables and constraints is determined by the type and number of subsystems modeled and the number of time steps incorporated in the control horizon.

The summary of methods given in the remainder of this section is very general. Various combinations of methods are used in practice. It should therefore be kept in mind that the approach and results of specific subclasses of methods and solvers may deviate from the ones discussed.

Within the framework of the current study, parts of the following sections have been published by Lobbrecht & Vredenberg (1995). The discussion of the methods applied is kept brief. Practical examples of the problem formulated have been worked out for the present research by Vredenberg (1996).

## 3.4.2  Network Programming

In general, networks are structures that can be described by arcs and nodes. Network Programming makes use of this special structure and therefore it can only be used when an optimization problem can be written in network form. However, not each network problem can be solved by NP.

It is important to note the difference between the term 'network' as used in literature on water-system simulation and in NP. In simulation it is common practice to divide a flow problem into arcs and nodes, in which the arcs represent flow elements and the nodes represent storage elements. In NP, flow elements and storage elements are both represented by arcs in the optimization problem. This phenomenon will be further described the remainder of this section and in Sec. 6.2.7.

## Modeling Characteristics

Network Programming poses the strongest limits on modeling of all methods discussed here. The limits on an NP problem make it easy to understand though, since there is a strong resemblance between the network representation of a water system and the graphical representation of the corresponding NP model (Fig. 3.2).

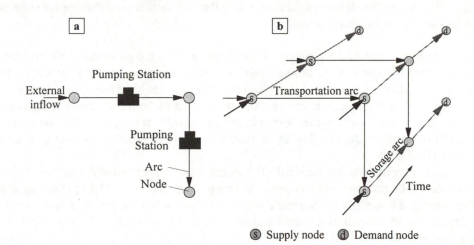

Fig. 3.2. Network model (a) and graphical representation of Network Programming (b).

As shown in Fig. 3.2, flows through structures are modeled by *transportation arcs*. Water stored in reservoirs is represented by *storage arcs*, since the NP model only allows flows. For that reason an NP-based model is also called a *network-flow model* in literature.

In applications of NP, network *supply nodes* and *demand nodes* are distinguished through which all inflows into and outflows from the system are modeled ('s' and 'd' in Fig. 3.2).

Network Programming requires a special structure of constraints. The constraints are defined as:

$$\overline{g}(\overline{x}, \overline{u}) = A \begin{bmatrix} \overline{x} \\ \overline{u} \end{bmatrix} - \overline{b} = \overline{0},\tag{3.2}$$

in which the constraint vector $\overline{g}(\overline{x}, \overline{u})$ contains the constraint functions $g_i(\overline{x}, \overline{u})$, $\overline{x}$ is the vector containing $m$ state variables, $\overline{u}$ is the vector containing $n$ control variables, $\overline{b}$ is the vector containing $l$ supplies and demands and $A$ is an $l \times (m+n)$-matrix. Matrix $A$ only consists of elements with values of -1, 0 and 1, where each column of $A$ has precisely one value of -1 and precisely one value of 1. $A$ is called the node-arc incidence matrix.

Every row in $A$ represents a node in the representation of the NP model, while every column represents an arc. The value -1 in a row indicates that an arc leaves the corresponding node, while the value 1 indicates that an arc arrives in the corresponding node. Every state and control variable in the model represents the flow through the corresponding arc.

In addition to the special structure required for the constraints, NP also requires the objective function to be convex and piece-wise linear. In NP the objective function is a total cost function. A cost-coefficient is required for the flow through each arc and the optimal solution is determined by minimizing the total cost.

A variation on the standard NP problem as described above is the generalized NP problem. In a generalized NP problem, the nonzero entries in the node-arc incidence matrix $A$ are no longer restricted to the values 1 and -1. This means that each arc has a multiplier at the node it leaves as well as at the node it arrives. Such multipliers make it possible to switch over from one source of flow to another source in the same problem e.g., to switch from water quantity flow to water quality flow or to switch from water quantity flow to generated electricity (Roos, 1987).

Another variation on the standard NP problem is an NP problem with *side constraints*. The standard NP problem only contains continuity equations. Modeling the physical relationships in a water system in general requires equations that do not fit in this network flow format. Additional side-constraint equations, can therefore be added to the network. As long as the number of side constraints is low in comparison to the continuity constraints, special solvers can be used that make use of the predominant network structure of matrix $A$ (e.g. Sun et al., 1995).

NP has been thoroughly described by Jensen & Barnes (1980) and Roos (1987). Considerations for solving the control problem as defined in the present study are discussed in Sec. 6.2.7.

### *Applications in Water-System Control*

Male & Soliman (1981) describe a mathematical programming procedure that can aid in the management of the Chicago River and Canal system. It involves formulating their water-control problem as an NP problem that can be solved using an efficient NP solver. It determines the minimum amount of water, diverted from Lake Michigan, necessary to maintain dissolved oxygen standards in the Chicago River and Canal System.

Kuczera (1989) shows that there are several multi-reservoir, multi-period Linear Programming formulations that can be transformed into NP problems. Network Programming is advantageous, mainly because general NP codes are readily available that are about 100 times faster than general LP codes. Network Programming therefore allows significantly larger multi-reservoir, multi-period problems to be solved than would be possible with LP. An NP formulation is presented for determining water assignments in a multi-reservoir system over a control horizon. The optimization approach provides the capability to handle shortfalls in supplies due to droughts, inflow requirements that may be violated during droughts and seasonal reservoir target volumes.

Chung et al. (1989) incorporate an NP problem into a simulation model for the California Aqueduct. They use the NP optimization problem to enhance the mass balance routing procedure of the simulation model.

Lobbrecht (1994b) uses an NP problem in combination with a simulation model, to determine the optimal control strategy for an regional water-resources system, describing water quantity only. The simulation model calculates storage levels while the optimum system flows are determined by solving an NP problem. The simulation model is basically used to calibrate the NP problem and to prevent the accumulation of errors caused by the linearizations used in the NP problem.

Nelen (1992) uses an NP problem to minimize combined sewer overflows in an urban drainage system during periods of high hydrological loads. The flow process through the sewer system is solely modeled by continuity constraints during a control horizon. The cost coefficients of the objective function are updated during successive calculations to incorporate nonlinear damage functions.

### Advantages

As can be seen in Fig. 3.2, the NP problem, in graphical terms, closely resembles the actual system. Furthermore, the constraints in the problem are simple continuity constraints. An NP problem is therefore more easily understood than problems based on other methods such as DP and NLP. During modeling this is a great advantage, since it is easy for the modeler to ensure that the problem has a feasible solution.

Computer codes to solve NP problems are readily available and most of these codes are very fast because they efficiently use the structure of the problem to determine the optimal solution. NP solvers are about 100 times faster than LP codes (Bertsekas & Tseng, 1988a, 1988b).

The order of dimensionality in NP is lower than $l \times (m+n)$ and will usually be more or less linear in the number of reservoirs, because the node-arc incidence matrix is extremely sparse.

### Disadvantages

The main disadvantages are associated with the accuracy of the NP problem. Constraints are not only required to be linear, but also should have a special continuity structure. Furthermore, the objective function has to be convex and piece-wise linear.

Many problems cannot be formulated as an NP problem without an unacceptable level of inaccuracy. Therefore, great care is required during modeling to ensure that the solution has a practical value.

### 3.4.3  Linear Programming

There is no doubt that Linear Programming is one of the most successful mathematical optimization methods. Hillier & Lieberman (1990) even rank Linear Programming among the most important scientific advances of the mid-twentieth century. In their book they describe the simplex method for solving LP problems. In recent years, the interior-point method has proven to be a competitive alternative to the simplex method for large problems, especially when matrix $A$ is sparse.

*Modeling Characteristics*

In Linear Programming, the objective function has the same restrictions as in NP. It has to be convex piece-wise linear in the state and control variables. In contrast to NP however, the constraints in LP are only limited by being linear, this enables much more flexibility in modeling. The constraints of an LP problem are still given by Eq 3.2, the elements of matrix $A$, however, are no longer restricted. It should be noted that NP is in fact a special case of LP.

Nonlinear relationships can be included in LP modeling through linearized approximations. Whether such linearizations are allowed, depends on the level of nonlinearity of the original physical processes in the system modeled.

Linear Programming is used both in deterministic and stochastic form (Yeh, 1985). The latter form can be used to model uncertainties in the system load e.g., in water-demand and precipitation predictions. The deterministic modeling approach does not incorporate uncertainties and assumes the hydrological-load predictions, if used, to be true. The stochastic approach incorporates uncertainties by taking into account that constraints have a certain probability of being true. In the stochastic approach a problem of dimensionality may arise. Stochastic LP is not further discussed here, LP problems referred to are of the deterministic type.

Linear Programming has been thoroughly described by Hillier & Lieberman (1990) who pay special attention to the simplex algorithm. In-depth discussions of interior point algorithms are given by Terlaky (1996).

*Applications in Water-System Control*

Crawley & Dandy (1993) use Linear Nonpreemptive Goal Programming for the Adelaide headworks drinking-water distribution system in South Australia. The planning and operational policies obtained aim at minimizing pumping costs, while ensuring system reliability by maintaining minimum target storage levels in the Adelaide reservoirs. Their reason for choosing LP is that fast commercial LP codes exist that can handle large problems. Network Programming could not be used since the structure of the problem did not permit solely continuity constraints. They see LP as an advantage since it enables future development of the model, including water-quality modeling.

Diba et al. (1995) use LP in the planned operation of large-scale water distribution systems. The LP problem will be used by operators to determine whether the existing mode of operation can be improved. Their model is also used as a training tool for operators to learn how to handle competing objectives.

*Advantages*

As for NP, there are several readily available solvers for LP. Although LP solvers are significantly slower than NP solvers, they are still fast when compared to solvers that implement other optimization methods. Most LP solvers support the use of the Mathematical Programming Standard (MPS) format of the input. By writing an LP problem in MPS format, it is possible to compare the performance of different solvers. This makes choosing the best solver for each type of problem straightforward. Moreover, when new solvers become available, they can be tested with relative ease and subsequently be incorporated if required.

The LP problem may not be as simple as the NP problem, but a larger variety of water-control problems can be solved using LP. Because LP requires no special structure for the constraints, modeling freedom is greater than in NP. Therefore, future extensions to a problem can be incorporated more easily than in NP.

Deterministic LP has a relatively low dimensionality when compared with nonlinear methods. Therefore, the problems which can be efficiently solved by using LP are much larger in size (number of subsystems and length of the control horizon) than by using nonlinear methods as DP and NLP.

*Disadvantages*

In LP, as in NP, the objective function is limited to being a convex piece-wise linear function of the state and control variables. Despite the greater modeling freedom in the constraints when compared to NP, the constraint functions still have to be linear, which can lead to problems when modeling highly nonlinear systems.

In comparison with NP, LP has the disadvantage that matrix $A$ is no longer guaranteed to be sparse. Maintaining the highest level of sparcity is the responsibility of the modeler. In general, in LP the order of dimensionality is significantly higher than in NP. Therefore LP can handle smaller problems than NP.

### 3.4.4 Successive Linear Programming

Successive Linear Programming can be considered an extension to LP. In LP, the linearizations used can lead to inaccurate or even invalid solutions. To avoid this problem, SLP uses an iterative solution approach, in which an LP problem is solved at each iteration.

### Modeling Characteristics

Successive Linear Programming is powerful for solving large nonlinear problems. Similar to the LP problem, an SLP problem uses linearized approximations of the nonlinear relationships. These approximations may occur in the objective function as well as in the constraints. After finding the optimal solution to the linearized problem, a new approximation to the nonlinear relationships is determined at the optimal values of system variables found. Then the updated LP problem is solved, etc. During the iterative process, the linearizations become more accurate. If applied well, the process converges and the final linearization point closely matches the final optimal solution.

Despite the great resemblance to LP, SLP is usually classified as NLP in literature. However, the problem solved at each iteration in the SLP procedure is a pure LP problem. It is the iterative approach that distinguishes SLP as an NLP method.

### Applications in Water-System Control

Martin (1987) describes a real-time application of a nonlinear problem for the operation of surface-water systems of the Lower Rio Grande System in Texas, USA. The problem is solved by flow routing and SLP. SLP is used to iteratively adjust daily reservoir releases to improve the water-system performance. In the method applied, the optimal solution of an LP problem is efficiently found by starting the search procedure from the feasible solution of a previous iteration loop.

### Advantages

In principle all advantages of LP are valid for SLP. However, the inherent inaccuracies resulting from linearizations in LP, are reduced to a large extent. The method becomes very powerful if, in successive iterations, the solution found in each previous iteration can be used as a starting point for the next iteration, this is called a *warm start*. The use of a warm start by a solver generally reduces the solving time dramatically.

This approach, combined with the high speed of LP solvers, makes SLP very competitive to pure NLP.

### Disadvantages

As in LP, for each problem in the iterative process, the objective function is limited to being a convex piece-wise linear function of the state and control variables in SLP. The objective function and the constraints have to be linear for each of the problems to solve. The order of dimensionality is slightly higher than in LP.

The intrinsic problem with SLP is that possibly highly nonlinear processes have to be linearized. Each iteration requires linearization and the optimization problem has again to be solved. This can take a lot of time if the processes modeled are highly nonlinear and the steps of the control horizon large in comparison to the time scale of the processes modeled.

To make the iteration converge to an optimal solution of the nonlinear problem, near-optimal values to linearize at, should be known with enough accuracy. This may require many calculations of the same time step.

A special form of SLP, without the majority of the disadvantages mentioned above is used in the present study and will be discussed in detail in Chapter 6.

### 3.4.5 Dynamic Programming

Many allocation and routing problems can be divided into a sequence of subproblems. In operational water control the control strategy can be considered a sequence of related decisions in time. Dynamic Programming provides a systematic procedure for determining the optimal combination of these decisions.

*Modeling Characteristics*

In Dynamic Programming, each time step of a control horizon is described as a separate subproblem. The subproblems are connected through a recursive relationship that is incorporated in the model. The recursive relationship describes the objective function for the time step under consideration ($t$) as a function of the objective function for the following time steps ($t+1, .. , T$; in which the control horizon $T$ is measured in discrete time steps).

The time steps indicated above, are called 'stages' in DP. Solving a DP problem, means first finding the optimal solution for the last stage. This depends on the system state at the end of time step $T$-1. The problem to be solved is then gradually enlarged by finding the current optimal solution for the preceding time step. This process continues until the original problem is entirely solved, using the known initial system state as a starting point (Hillier & Lieberman, 1990).

Two types of Dynamic Programming are distinguished. The first version requires all the state variables to be discrete and is called discrete DP. The second type uses continuous state variables and is called continuous DP. The latter is not discussed further here.

Like in LP, stochastic processes can be included in DP modeling. These are not discussed any further here.

Dynamic Programming should be seen as a general approach to problem solving, there are no standard solvers that can be applied. Any solvers that have been developed are very problem-specific.

*Applications in Water-System Control*

Yeh (1985) attributes the popularity and success of DP to the fact that the nonlinear and stochastic features which characterize a large number of water resources problems, can be incorporated in a DP formulation. Moreover, DP has the advantage of effectively decomposing highly complex problems involving a large number of variables, into a series of subproblems which are solved recursively.

Constraints that restrict the solution space, are advantageous in discrete DP, because they reduce the amount of computations necessary to solve the problem, this in contrast to other optimization methods such as NP, LP and NLP. The reason for this advantage is that for each possible system state $\bar{x}(t)$, an optimal control $\bar{u}(t)$ has to be calculated. By restricting the state space, the number of possible system states that have to be taken into account, is reduced. Similarly, restricting the solution space reduces computation time, since fewer possible solutions have to be taken into account. This feature of DP allows the optimization process to start by solving the problem using a coarse discretization and then, when the optimal solution is nearly known, solving the problem again using a fine discretization close to the near optimal solution.

When applied to multiple-reservoir systems, the usefulness of DP is limited by the 'curse of dimensionality'. In DP the order of dimensionality is exponential to the number of state variables, e.g. reservoirs. In discrete DP all possible combinations of values of the state variables have to be taken into account at each stage. Adding a state variable, multiplies the number of possible combinations by the number of possible values for the new state variable.

Crawley & Dandy (1993) rejected DP for the Adelaide headworks system in South Australia because using DP would involve combining reservoirs to reduce the problem of dimensionality. This would result in loss of data that are necessary to determine the desirable set of reservoir storage levels at various times throughout the system.

Mariño & Mohammadi (1983) use a combination of LP and DP to maximize the annual hydro-electric energy and annual municipal and industrial water releases from a system of reservoirs. LP problems are solved for each stage of one month in the DP optimization. The objective of these LP problems is to minimize the releases from the reservoirs, given a set of constraints. These constraints are contracted water and energy supplies, continuity equations and hydrological process descriptions.

### Advantages

Dynamic Programming has some advantages over NP and LP. In particular, DP enables nonlinear and nonconvex objective and constraint functions to be used. This means that the complex nature of regional water systems can be incorporated better than in NP and LP.

A useful feature of discrete DP is that strict limits on the solution space reduce the number of system-state evaluations. Therefore, limits incorporated into a problem, improve the performance of DP. In contrast, in NP, LP and NLP problems, extra limits on the solution space are essentially extra constraints that result in extra search directions which reduces their performance.

### Disadvantages

The most significant disadvantage of DP is the curse of dimensionality. At each stage of the discrete DP solution process, several alternative solutions have to be generated and each of these solutions has to be stored in order to recursively determine the best combination of these solutions. The number of alternative solutions is exponential in the number of state variables.

For methods such as NP and LP the order of dimensionality has a linear relationship to the number of reservoirs and therefore larger problems can be solved by using NP and LP than by using DP.

The solution process in DP is computationally slow. This is due to the fact that each stage of the optimal control strategy has to be determined explicitly for all possible system states. Therefore, only relatively small problems are solvable.

A DP problem is difficult to understand. This can be a major problem during modeling and implementation. Moreover, the lack of a standard mathematical formulation of 'the' DP problem, has prevented the development of a general DP solver.

Dynamic Programming has no restrictions on problem formulation. However, it is not guaranteed to find a global optimum if the objective and/or constraint functions are nonconvex.

### 3.4.6 Nonlinear Programming

The Nonlinear Programming method involves a large variety of problems. In fact every problem that is not an LP problem, is an NLP problem. Strictly speaking DP could be seen as a subclass of NLP. NLP and DP are described separately, however, because the way in which a solution to the control problem is found differs essentially. The application of NLP to the optimization problem defined, has been studied by Vredenberg (1996).

*Modeling Characteristics*

Throughout the years several solution methods have been developed to solve some important types of NLP problems. Each of these methods, solves the problem by iteratively improving the value of the objective function, starting from a feasible starting point.

At each iteration point a search direction is determined which improves the value of the objective-function. A line search along the search direction is then performed to find a better solution than the current one. The state and control vectors corresponding to the improved function value, are subsequently taken to be the new iteration point.

Theoretically, there are no restrictions on NLP problems, in practice, however, this is not the case. Solvers are generally solely applicable for certain subclasses of problems, requiring special properties of the functions in the NLP problem.

Three major classes of NLP solution methods can be distinguished, each of which makes certain assumptions about the objective and constraint functions. The first class uses constraint and objective functions to determine the search direction. The second class also uses first-order partial derivatives of the constraint and objective functions. The third class, in addition, uses second-order partial derivatives.

Each of these classes requires the objective function and the constraint functions to be continuous. The second and third class, in addition, require the first- and second-order partial derivatives to be continuous respectively.

While the restrictions posed on the problem increase from the first to the third class, solution times decrease. This is because more information on the right search direction is available which can be used during the search process. However, the memory requirements increase, since the partial derivatives have to be stored.

### Applications in Water-System Control

Wide use of NLP has been hampered for a long time, because the process of optimization is slow and the order of dimensionality is large in comparison to NP and LP (Yeh, 1985). Moreover, the mathematics required are considerably more difficult than those required for NP or LP.

In the USA, the NLP modeling in operational water control is rapidly becoming common practice. This is a result of the great modeling freedom that NLP allows, in contrast to NP and LP.

Authors who report on using NLP often state that optimization problems would have to be simplified too much if linear methods are to be used. NLP has the advantage that the exact relationships as used in simulation models can be incorporated without further simplification. On the other hand, NLP modeling can generate very complex optimization problems.

Mays (1989) discusses an alternative to solely using simulation or solely using NLP in his paper 'Simulation versus Optimization: Why not both?'. He describes three examples of coupled simulation models and NLP problems, where a reduced NLP problem is solved in conjunction with a simulation model. This method enables NLP to be used and prevents the accumulation of inaccuracies in the NLP problem.

Pezeshk et al. (1994) note that in many cases reality is simplified in order to avoid nonlinearities and nonconvexities. They show that LP is sometimes used beyond its limits of applicability, to solve basically nonlinear optimization problems. They describe an NLP problem to minimize pumping costs for the operation of a well field and a main drinking-water distribution system.

### Advantages

Nonlinearities that characterize regional water systems can be modeled in more detail using NLP than by NP, LP and DP. The NLP method gives a more accurate optimal solution than the NP, LP and DP methods. In contrast to DP, readily available solvers for specific classes of NLP problems exist.

*Disadvantages*

A major disadvantage of NLP is the complexity of the optimization problem to define and solution method. This makes application difficult. In LP, the linearization of basically nonlinear processes may be difficult when accuracy is required, in NLP it is the modeling itself. One of the major problems is to determine continuous physical relationships and, depending on the method used, also continuous derivatives of these relationships. The determination of the model relationships, generally requires approximations of the physical relationships.

The large order of dimensionality of the NLP problem and the often long solution times, limits the size of the problems that can be solved. This means that several assumptions may have to be made during modeling, which introduces the risk of oversimplifying. Furthermore, the solution process can slow down dramatically, close to the optimum.

Most algorithms implemented in NLP solvers have the inherent disadvantage that first a feasible point in the solution space has to be found before real problem solving can start. This is especially a problem for complex problems with strongly limited variables. In addition, the feasibility of the solution can get lost during optimization. Special measures are built into solvers to return to a feasible point in the solution space again, but this generally requires many extra iterations.

As is the case with DP, the optimum found is not always guaranteed to be the global optimum. A global optimum is only guaranteed if convex objective and constraint functions are used.

### 3.4.7 Summary

All mathematical optimization methods discussed in the preceding subsections have advantages and disadvantages. Advantages of the NP and LP methods are the possible large sizes of the water systems which can be modeled and the solution speed in comparison with DP and NLP. However, NP and LP are less accurate than DP and NLP.

In the USA most work has focused on DP-related methods traditionally, but more recently on NLP-related methods. The choice for these nonlinear methods is to a large extent a result of the important role of hydropower generation that is described by strongly nonlinear functions. Maximization of hydropower is often one of the most important objectives in the control of the water systems in the USA. According to the literature found, the functions cannot be linearized with enough accuracy to apply linear methods. It should be noted though, that the general control horizons (months) used for these water-release-planning problems are much longer than the control horizons used in the present study (days).

In the Netherlands, the emphasis has traditionally been more on NP- and LP-related methods, mainly because the descriptions of the important processes in the water systems of the Netherlands are not predominantly described by nonlinear relationships as is the case in the USA (hydropower). In addition, the operational control problems in the Netherlands require models with a large number of subsystems and relatively small time steps to incorporate the system-load prediction accurately. Furthermore, the control horizons involved in operational control are relatively short, which enables continuous correction of inaccuracies in the control strategies determined.

Some of the disadvantages of the nonlinear methods of DP and NLP can be reduced considerably by applying SLP. Although the underlying processes of NLP and SLP are different, SLP in fact can be considered a special form of NLP, since nonlinearities can be modeled in stages.

Table 3.1 summarizes some key properties of the optimization methods described. The global optimum property shown in last column of the table should be considered with respect to the optimization problem as defined.

Table 3.1. General characteristics of the mathematical optimization methods described.

|  | Solvable problem size | Solution speed | Model accuracy | Model complexity | Solvers available | Global optimum |
|---|---|---|---|---|---|---|
| NP | Very large | Very fast | Low | Very low | Yes | Yes |
| LP | Large | Fast | Low to moderate | Moderate | Yes | Yes |
| SLP | Large | Moderate to fast | Moderate to high | Moderate | Yes | Yes, for each iteration |
| DP | Very small | Very slow | High | High to very high | No | If functions are convex |
| NLP | Moderate | Slow | High | High to very high | Yes | If functions are convex |

## 3.5    Reasons for Choosing Successive Linear Programming

Several criteria are relevant and affect the choice of mathematical optimization method to be used in the current study:
•    objectives have to be accurately incorporated in the optimization problem;
•    system characteristics have to be modeled with sufficient accuracy;
•    alternative water-system arrangements should be modeled easily;
•    the optimal solution has to be determined much faster than real-time;
•    finding the global optimum of the problem built should be guaranteed;
•    to enable detailed modeling, large problems should be solvable.

On the basis of these criteria and the results of the survey given in Sec. 3.4, deterministic Successive Linear Programming was chosen as being the most promising method to solve the mathematical problem formulated.

Successive Linear Programming enables modeling of nonlinearities and therefore, in principle, nonlinear objectives and nonlinear water-system behavior can be described with sufficient accuracy. To avoid instable solutions and to speed up the search process of the SLP method in general, in the application described in Chapter 6, the objectives will be linearized explicitly and used during successive iterations without modification.

The accuracy gained with SLP may not be as high as with pure NLP, but it is questionable whether accuracy improvement in the control strategy found is very meaningful. As will become clear in Sec. 5.3, the hydrological load which is included in the optimization problem is basically not very accurate.

Similar to NP, LP and NLP, an SLP-based model can be changed or extended relatively easily.

Especially in real-time control, the speed of finding the solution to the optimization problem is of great importance when choosing a method. In fast-reacting water systems where critical discharge times are of the order of minutes (sewer systems) to hours (rural drainage systems), the automatic control system or the responsible operator has to act fast. Therefore, the time to solve the problem should at least be in the same order of magnitude and preferably shorter.

To enhance the SLP method, a special technique, here called *forward estimating* is developed in the present study (Sec. 6.2.2 and 7.1.6), which makes it almost as fast as LP and almost as accurate as NLP. SLP guarantees finding the global optimum of the LP problem solved for each stage.

The size of problems that can be solved by SLP equals that of LP, which is large, generally one order of magnitude larger than NLP, when the same computer resources are available.

Not an explicit advantage of (S)LP, but still of importance to solving the problem, is the fact that several readily available solvers can be used. These solvers have different characteristics and some of them make effective use of the structure of the optimization problem. Moreover, when modeling, the problem structure can be easily adapted to the special features of the solver, increasing the overall solving speed.

As in all mathematical programming methods described in this chapter, discontinuities in water-system descriptions are very difficult to incorporate in SLP. Furthermore, some highly nonlinear processes are sometimes hard to capture by means of linearization, producing unstable results in stage-wise linearization.

These problems can be reduced by combining mathematical optimization with simulation. In that case, the optimization problem can include simplifications, while the simulation model accurately describes the processes in the water system. This approach, which resembles the one of Mays (1989), is used in the present study. A general introduction to the method is presented below.

## 3.6    Simultaneous Simulation and Optimization

A method combining mathematical optimization with simulation has been developed to solve the operational control problem. Optimization is applied to determine control actions some time ahead, while simulation keeps the optimization process accurate. Simulation and optimization are implemented by two separate modules in the Strategy Resolver, (Sec. 2.4.2): an optimization module and a simulation module. In this context the term *module* is used to indicate part of a computer program. The optimization module solves the 'optimization problem', while the simulation module runs the 'simulation model'.

Figure 3.3 shows the interactions of simulation and optimization. It should be realized that application of the Decision Support System developed in this thesis, is only demonstrated in analysis mode, which can be considered off-line use by the control and tactical organizational levels in a water authority.

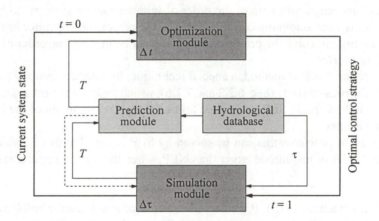

*Fig. 3.3. Simultaneous simulation and optimization as implemented in the Strategy Resolver.*

A simulation model incorporates a description of the nonlinear relationships of processes in the water system. The optimization problem contains a simplified and linearized description of these processes. In simulation, the response of the water system is calculated on the basis of hydrological data. Time-series calculations are made for both long periods and for specific periods of extreme hydrological conditions, to determine the behavior of the water system under average and high hydrological loads, respectively.

The optimization module determines the optimal control strategy for the water system, taking into account the objectives set for interests during the control horizon. The control horizon contains a discrete number of $T$ time steps of size $\Delta t$. The length of the control horizon may range from a few hours to several days or weeks.

The input to the optimization module consists of data on the system state at simulation time $\tau$ and a prediction of the system load during the control horizon. This prediction is determined by the *prediction module*.

The prediction module basically determines the hydrological load and the loads to the water system which are included in the simulation model, but not in the formulation of the optimization problem. For the latter purpose, the prediction module makes use of the simulation module.

The prediction module can also be used if only simulation is applied, to determine a control strategy by trial and error (Sec. 2.4.2). This is represented by the dotted line in Fig. 3.3 but will not be discussed further.

At each simulation time step $\tau$, the simulation module receives the optimal control strategy from the optimization module and dynamically calculates the system state corresponding to the basic setpoints of time $t = 1$ (Sec. 2.3.1). Examples of variables that determine the system state are: surface-water levels, groundwater levels and concentrations of water quality variables.

Simulation is only required for the set of control actions associated with optimization step $t = 1$, all subsequent actions when $t > 1$ are ignored. For every step, the hydrological variables are predicted again. By doing so, the control actions associated with $t > 1$ become more accurate in the next loop, where they are determined again on the basis of updated predictions (Fig. 3.4).

The system state used by the optimization module, is updated at every optimization time step, to correct for the inaccuracies that are the result of its simplified form. To enable strategy determination for long control horizons, in principle, many optimization steps would be required. To keep an optimization problem within workable size, the time-step size used in optimization ($\Delta t$) can be chosen larger than the one used in simulation ($\Delta \tau$). However, for each model built, a sensitivity analysis should be performed to check whether such a difference between $\Delta t$ and $\Delta \tau$ is allowed.

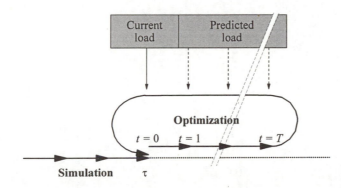

*Fig. 3.4. Current and predicted hydrological load.*

## 3.7    Concluding Remarks

This chapter presented the general formulation of the mathematical optimization problem for finding the optimal control strategy for the general water system and showed the methods which can be used to solve that problem. The mathematical programming method of Successive Linear Programming has been identified as the most promising method, since it can incorporate nonlinearities with a rather high accuracy, as long as the water-system processes incorporated in the optimization problem are not highly nonlinear. The method is fast in comparison to nonlinear methods such as Dynamic Programming and Nonlinear Programming. The speed in finding solution to optimization problems, is demonstrated in the case studies and performance analyses of Chapter 7.

To model large water systems in detail, a method of simultaneous simulation and optimization has been especially developed. The nonlinear relationships of the processes in the water system described by the simulation module are presented in Chapter 4. Linearized descriptions of processes, which can be handled by the optimization module, are presented Chapter 6.

To be able to dynamically reallocate capacities in the water system and to anticipate hydrological loads on the water system, predictions of hydrological variables are sometimes needed. Chapter 5 describes how to determine the hydrological load on a water system.

# 4   Simulation of Regional Water Systems

## 4.1   Introduction

Regional water systems consist of hydrological units, which interact via natural and artificial water flow paths. The current chapter describes regional water systems in which control has a major impact. Several elements of water systems are described in detail, to show state-of-the-art modeling, that is required to solve the general control problem of the water system.

The introduction of control elements in a water system changes the original hydrological system and splits it up into separate *areas*, each consisting of one or more subsystems. The classification shown in Fig. 1.3 (Page 2) is used to define the various subsystems, i.e.:
- urban drainage,
- urban surface water,
- urban groundwater,
- rural surface water,
- rural groundwater.

Subsystems interact via discharge structures, called *flow elements*. Flow elements can be divided into controllable flow elements, here called *regulating structures*, fixed flow elements, here called *fixed structures* and *free flow elements*. A regulating structure is defined as a structure that is operated regularly, either manually or automatically, adjusting the flow capacity. Examples of regulating structures are pumping stations and controllable inlets. An example of a fixed structure is a fixed weir. Examples of free flow elements are: canals and regional groundwater flow.

A water system can be described by subsystems and flow elements. The number of subsystems and flow elements that have to be considered, depends on the amount of detail needed to describe the corresponding processes.

The hydrology of a particular water system generally cannot be separated completely from surrounding water systems. In the description of a water system this forms a special point of interest. In this thesis, interactions with neighboring systems are included in the water-system description itself, by formulating the boundary conditions that influence or are influenced by the water system at the *water-system boundary*.

Figure 4.1 shows a typical regional water system, including various elements mentioned above.

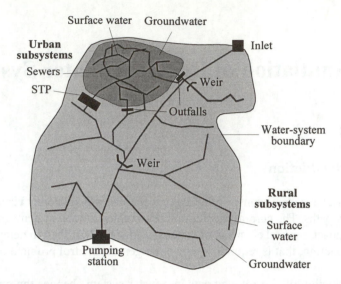

*Fig. 4.1. Typical regional water system: urban and rural areas with subsystems and flow elements.*

## 4.2   Hydrology of Subsystems

### 4.2.1  Surface Types

This section discusses various types of surfaces that can be distinguished in regional water systems and the associated runoff processes that are of importance to regional water-system control. The subsystems of an urban area can incorporate rapid and slow runoff processes, such as rapid runoff via sewer systems and slow runoff via the permeable subsurface of green belts. Here, such runoff processes are classified on the basis of their specific characteristics and not on the basis of their geographical location in the water system.

The following surface types affect runoff (Fig. 4.2):
*   pervious,
*   semi-impervious,
*   impervious.

***Pervious Surface***

Pervious surfaces are predominantly found in rural areas and green belts in urban areas. Precipitation $P$ falls on the soil and vegetation. Part of the precipitation is intercepted by local depressions and plant leafs. Most of it, however, infiltrates into the soil.

From the soil or other wet surfaces, water may evaporate directly. The combined evaporation from depressions, soil and plant leafs is called evapotranspiration, indicated by $E_t$. If water can flow unobstructed over the land surface, water from depressions may be discharged directly to the surface-water subsystem.

A special type of pervious surface is found in infiltration systems in urban areas. These surfaces are created specifically very pervious to enable the runoff from roofs, roads and parking places to infiltrate into the subsurface. In general, such areas are created to keep precipitation as much as possible in the urban environment, thus preventing a loss from the urban subsystem by allowing recharge of the groundwater.

### Semi-Impervious Surface

In general, semi-impervious surfaces are found in urban areas. This surface type comprises brick-paved roads, parking areas and other pavements that allow infiltration of precipitation into the soil. Paved areas that allow infiltration are called 'open pavements'. Such pavements can be especially constructed to recharge the groundwater.

Precipitation that falls on semi-impervious surfaces first accumulates in depressions. From there, it may either evaporate ($E_0$), flow to surface waters or infiltrate into the soil. In general, the infiltration rate is so high, that precipitation of a low intensity only infiltrates. Only when large amounts of precipitation fall on semi-impervious surfaces, part of it flows into sewers or runs off to surface waters (Van de Ven, 1989).

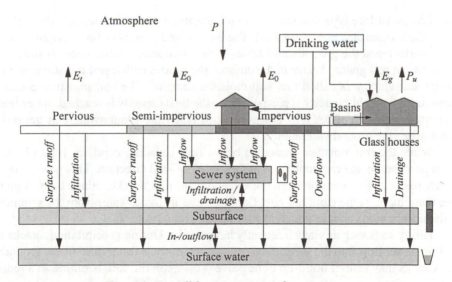

Fig. 4.2. Runoff from various surface types.

### Impervious Surface

Roofs, bitumen-covered streets and parking lots are impervious surfaces. Precipitation on such surfaces accumulates temporarily in depressions and forms puddles. This water may evaporate ($E_0$) directly. Depending on the quantity of precipitation and its intensity, excess water flows either to neighboring surface waters or into sewer systems.

Wastewater from homes and industry is discharged as sewage water into sewer systems. A small, and fortunately continuously decreasing amount of waste water is discharged directly onto surface waters. With respect to water quantity, this amount can be neglected. With respect to water quality, this direct discharge can be significant because of its polluting character.

A special type of impervious surface is the surface occupied by glasshouses. The total area covered by glasshouses is generally limited, but concentrated at specific locations in water systems. Especially in polders, such glasshouse areas can play an important role in water-system control.

## 4.2.2 Subsystem Interactions

This section describes the interactions between the several subsystems distinguished in Sec. 4.1. The flow processes in the unsaturated zone are described separately in Sec. 4.2.3.

### Subsurface

Depending on surface type and precipitation intensity, water infiltrates into the soil, where it enters the *unsaturated zone* (Fig. 4.3). The unsaturated zone is defined the zone between the soil surface and the groundwater table. In the unsaturated zone, water is stored in the pores between soil grains. Water in the unsaturated zone is called soil moisture; its ratio to the total soil volume is called the 'soil moisture content'. The soil moisture content can increase until the field capacity is reached. Once the field capacity is reached, water from the unsaturated zone percolates ($P_r$) downwards, to reach the groundwater. In general, this process raises the groundwater table.

When the soil moisture content is below field capacity, capillary forces in the soil prevent percolation and can contribute to flow in an upward direction. This process is called capillary rise ($C_r$). Evaporation from the soil and plant-root uptake reduce the soil moisture content and increase the capillary rise. Capillary rise abstracts water from the groundwater into the unsaturated zone, in general lowering the groundwater table.

Heavy soils such as clays frequently have cracks. During precipitation, cracks in the soil allow faster flow into the subsurface and, consequently, a faster raise of the groundwater table. Cracks may restrict build-up of negative pressure in the soil, resulting in a reduction of capillary rise.

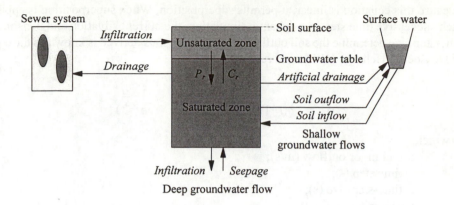

*Fig. 4.3. Subsurface flows.*

Two other processes contribute to inflow into the subsurface: infiltration from semi-impervious surfaces and infiltration from leaking sewer systems.

The zone below the groundwater table is called the *saturated zone*. In this zone, the pores between the soil grains are filled entirely with water. All flows to and from this zone are called *groundwater flows*. Two types of groundwater flow are distinguished: *shallow groundwater flow* and *deep groundwater flow*.

In general, shallow flow processes take place in the phreatic groundwater aquifer, which are the result of the hydraulic gradient of the groundwater table. Shallow groundwater outflow can be accelerated by subsurface drains.

The groundwater table may fall as a result of capillary rise and infiltration into deeper layers, especially in summer. As soon as the groundwater table falls below the level of surrounding surface water, inflow from surface water into the soil starts. In general, this inflow is restricted by the imperviousness of the canal beds and banks.

A special form of shallow groundwater outflow is the draining effect of leaking sewers, when the sewer system is located below the groundwater table.

The generalized flow equation describing the outflow from the soil to the surface water, the inflow from the surface water into the soil, or the outflow from drains is:

$$q = \frac{\Delta h}{S_r},$$

(4.1)

in which:

$q$      : flow into or out of the soil (m/s);
$\Delta h$    : difference between representative groundwater level and surface-water level (m);
$S_r$     : soil resistance to soil inflow, soil outflow, flow via drains (s).

Because this equation is linear, it permits superposition. When superposition is applied for each successive time step in a dynamic simulation, the current situation and the simulation time-step size determine the soil outflow. The equation thus derived is known as the equation of De Zeeuw-Hellinga (1958) and reads:

$$q(t) = q(t-1) \, e^{-\frac{t_s}{S_r}} + P(t)(1 - e^{-\frac{t_s}{S_r}}), \tag{4.2}$$

in which:
$q$      : soil in- or outflow (m/s);
$t$      : time step (-);
$t_s$     : time-step size (s);
$P$     : precipitation (m/s).

The equation assumes horizontal laminar flow. In addition to determining horizontal flow, the equation is used successfully to determine the outflow from soils in hilly areas, which have strong vertical components (Grotentraast, 1992).

Groundwater-pressure differences on a regional scale result in deep groundwater flow between deep aquifers and phreatic water, e.g.: infiltration towards lower aquifers and seepage from lower aquifers. This phenomenon is very common in polder areas.

    In practice, shallow and deep groundwater flow can exist together in regional water systems, depending on local soil characteristics, drainage possibilities and soil structure.

### Sewer Systems

Three major types of sewer systems are distinguished (Fig. 4.4):
1.     combined sewer systems (CS);
2.     separate sewer systems (SS);
3.     improved separate sewer systems (IS).

The combined sewer system (1) discharges sewage water together with excess precipitation. The sewage water produced by homes and industry is called dry-weather flow (DWF). The combined sewer system collects both DWF and excess precipitation. A combined sewer system has both a discharge and a storage function. DWF and excess precipitation is discharged to sewage treatment plants (STPs), where the water is treated and most pollutants are removed. The hydraulic capacity of an STP is limited, e.g. to 0.7 mm/hour, which is related to the area discharging to the sewer system. The effluent of STPs is discharged to the surface water.

    The storage capacity of a combined sewer system is limited, e.g. to 7 or 9 mm. Outfalls are constructed at specific locations to prevent streets and adjacent areas from being flooded by polluted water when the storage capacity is reached in part of or in the entire sewer system. This overflow is called combined sewer overflow (CSO). Some outfalls can be controlled, but most of them are fixed. In exceptional situations, the discharge capacity of the sewer system to the outfalls may become too low. Sewers flooding streets may be the undesirable result of such undercapacities.

The separate sewer system (2) involves separate sewers for DWF and rainwater. Wastewater from homes and industry is discharged by special dry-weather sewers. The discharge capacity of these dry-weather sewers can be much lower than that of the combined sewer system. All sewage water of these dry-weather sewers is discharged to STPs.

Excess precipitation water from impervious or semi-impervious surfaces such as roofs, streets and parking lots is discharged to rainwater sewers. Rainwater sewers discharge their water onto the surface water, called separate sewer discharge here (SSD).

The separate sewer system prevents discharge of waste water to the surface water, but the quality of discharges to the surface water is generally still poor. This is due to the fact that the catchment areas from where precipitation is collected to flow into the sewers, can be rather polluted. Furthermore, mistakes in connecting homes to rainwater sewers are a continuous source of surface-water pollution.

The improved separate sewer system (3) combines the advantages of combined sewer systems and separate sewer systems. It consists of two connected sewers systems for DWF and rainwater. The dry-weather sewer discharges to an STP where the water is treated.

Structures connecting the two sewer systems enable internal overflow of excess precipitation from the rainwater sewers to the dry-weather sewers. Once the dry-weather sewers have reached their maximum storage capacity, the connecting structures prevent further internal overflow. Then, the rainwater sewers have to store all further precipitation until their storage capacity is reached as well.

Once the storage capacity of the rainwater sewers is reached they overflow to the surface water, as happens in the combined sewer system. This overflow is called improved-sewer-system overflow here (ISO).

In comparison to the combined sewer system, the improved separate sewer system has the advantage that its overflows are far less polluted. The advantage over the separate sewer system is that rainwater initially polluted by roofs, streets and parking lots is flushed to the

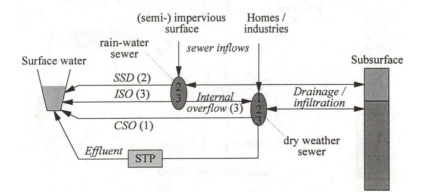

*Fig. 4.4. Sewer-system flows.*

dry-weather sewers, from where it is discharged to an STP. If applied well, surface waters are thus less polluted.

All types of sewer systems may have flow interactions with the surrounding soil as a result of sewers leaking. Depending on the sewer height in comparison to the groundwater table, infiltration into the soil or drainage from the soil may occur.

### Surface Water

The surface-water subsystem collects runoff as well as several forms of discharge as described above (Fig. 4.5). Surface water temporarily stores discharged water, which may also originate from other surface waters.

To discharge water or letting it in, various flow structures are used in practice, such as pumping stations, weirs, sluices, etc. These structures and the way in which they are controlled, are described in Sec. 4.3.

Surface water may also interact with the environment as a result of deep groundwater flows. The interaction depends on pressure differences between the surface-water level and the water level in the phreatic aquifer or the pressure height in the deep groundwater. Both infiltration from or seepage to surface water occur and the direction of flow may change during a year.

Precipitation to and evaporation from surface water in general has a limited impact on water-system control. These quantities are generally small in comparison to surface runoff or discharge from the subsurface and sewer systems.

In horticulture, good-quality water is extremely important because pollutants in irrigation water affect the root uptake by plants. In low polder areas saline seepage may occur, which results in surface water having a rather high salt concentration, especially in summer.

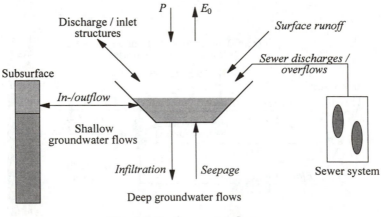

Fig. 4.5. Surface-water flows.

For that reason, glasshouse horticulture strongly depends on alternative water sources. Alternatives are drinking water or rainwater collected in rainwater basins. During excessive rainfall these rainwater basins sometimes overflow. The total amount of water withdrawn from a water system in which glasshouses are present, also called plant uptake ($P_u$), can affect the overall water balance. For example in tomato production, the total plant uptake may be as high as 50 mm per year, which is abstracted from the water system on harvesting. Evaporation in glasshouses ($E_g$) can be extremely high; 5 mm per day is no exception in summer.

Surface-water quality can generally only be influenced indirectly by regulating structures. With respect to water-quality development, the following differential equation and subsequent relationships can be derived for continuity in a surface-water subsystem for a specific pollutant:

$$\frac{dVc}{dt} = V\frac{dc}{dt} + c\frac{dV}{dt} = \sum (c_{in}Q_{in} - c\sum Q_{out}) + V\sum R \iff$$

$$V\frac{dc}{dt} = \sum (c_{in}Q_{in} - c\sum Q_{out}) - c(\sum Q_{in} - \sum Q_{out}) + V\sum R \iff \quad (4.3)$$

$$\frac{dc}{dt} = \frac{\sum c_{in}Q_{in} - c\sum Q_{in}}{V} + \sum R,$$

in which:
$c$        : pollutant concentration in surface-water subsystem (g/m³);
$V$        : volume of surface-water subsystem (m³);
$t$        : time (s);
$c_{in}$      : pollutant concentration in inflow into surface-water subsystem (g/m³);
$Q_{in}$      : inflow into subsystem (m³/s);
$Q_{out}$     : outflow from subsystem (m³/s);
$R$        : reactions (g m⁻³ s⁻¹).

The water-quality relationships described by Eq. 4.3 have also been used with success by Benoist et al. (1997). In this thesis, water quality focuses mainly on pollutants which restrict satisfying the requirements of interests. Of all the possible processes which can be described by Eq. 4.3, only the conservative and decay-related ones are taken into consideration. Conservative processes impose term $R$ to equal zero. For decay, the following equation applies:

$$R = -D_r c, \quad\quad (4.4)$$

in which:
$D_r$       : decay rate (s⁻¹).

Using this relationship, the following numerical equation for water-quality development can be determined on the basis of in Eq. 4.3:

$$c(t) = (1 - D_r \Delta t) c(t-1) + \frac{\sum c_{in} Q_{in} - c \sum Q_{in}}{V} \Delta t . \tag{4.5}$$

This equation requires $D_r$ to be zero for conservative substances.

### 4.2.3  Unsaturated Flows

Flows in the unsaturated zone of the soil have been studied and reported on by several authors (e.g. Feddes et al., 1988 and de Laat, 1980). These studies generally have an agricultural focus. How much water is available in the soil throughout the growing season is a prime concern to determine the growing potential for crops. In general, the time scale used in the studies is considerably longer than the one used in this thesis (e.g. months versus several hours/days).

Here, the focus is on the flow processes from and to the soil from a hydrological point of view. The purposes are:
*      to predict the actual moisture content in the root zone;
*      to predict the water quantity flowing out of the soil into the surface water or vice versa.

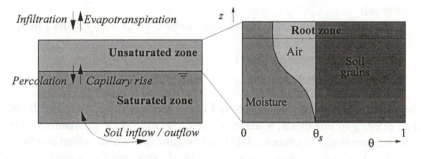

*Fig. 4.6. Unsaturated zone: terminology.*

Figure 4.6 shows the terms used in relation to unsaturated flow. In the root zone, plant roots take up water from the soil. The height of this zone is determined by the type of crop, its growing stage, soil texture and the groundwater level in spring. Flow processes to and from the saturated zone have been described in Sec. 4.2.2.

*Soil Moisture Content*

The soil moisture content and the water pressure in the soil determine whether plant roots are able to take up water from the soil. The soil moisture content varies with the location in the unsaturated zone. Just above the groundwater table, the soil is almost saturated and therefore, the moisture content at that level is called 'saturation moisture content' $\theta_s$. In an upward direction, the moisture content reduces, as is shown by the curve in Fig. 4.6. It should be mentioned that the exact shape of $\theta(z)$ depends on whether the groundwater table is falling or rising.

At the groundwater table the water pressure equals the atmospheric pressure. Above the groundwater table, the pressure is negative. This pressure, expressed in water column units, is called 'matric pressure' $h_m$. The development of the matric pressure in time depends on:

- the abstraction of moisture from the soil by plant roots, expressed in the sink term $S$;
- evaporation from soil with no vegetation;
- the location in the soil;
- depth of the groundwater table below the soil surface.

The relationship between the variables mentioned is given by Darcy's law (Eq. 4.6) and the continuity equation (Eq. 4.7) for unsaturated flow. The equations assume vertical flow and a homogeneous soil.

$$q = -K(h_m) \left( \frac{\partial h_m}{\partial z} + 1 \right) \tag{4.6}$$

$$\frac{\partial \theta}{\partial t} = -\frac{\partial q}{\partial z} - S(h_m) \tag{4.7}$$

in which:
$q$   : volumetric flux (cm/d);
$K$   : unsaturated hydraulic conductivity (cm/d);
$h_m$   : matric pressure (cm);
$z$   : soil depth, measured from the surface, with upward positive direction (cm);
$\theta$   : soil moisture content (-);
$t$   : time (d);
$S$   : sink term (d$^{-1}$).

Combination of the two equations, yields the Richards equation (Feddes et al., 1988):

$$\frac{\partial h_m}{\partial t} = \frac{1}{C(h_m)} \frac{\partial}{\partial z} \left[ K(h_m) \left( \frac{\partial h_m}{\partial z} + 1 \right) \right] - \frac{S(h_m)}{C(h_m)}, \tag{4.8}$$

in which:

$$C(h_m) = \frac{\partial \theta}{\partial h_m}. \tag{4.9}$$

In Eq. 4.9 $C(h_m)$ represents the soil moisture capacity in $cm^{-1}$, or the slope of the soil-moisture retention curve. The formulation of Eq. 4.8 is valid for the entire flow region of the soil-moisture retention curve, including the saturated zone. A reason for expressing the relationship in terms of matric pressure $h_m$ is that this is a continuous variable over the entire height of the soil profile, even when layered soils are considered.

Several numerical methods have been developed to solve Eq. 4.8. Examples of computer applications are SWATRE (Belmans et al., 1983) and MUST (De Laat, 1980).

Van Genuchten (1980) developed a mathematical relationship between soil moisture content and matric pressure which is now commonly used (Eq. 4.10), and an equation describing the hydraulic conductivity in unsaturated soil (Eq. 4.11):

$$\theta = \theta_r + \frac{\theta_s - \theta_r}{(1 + |\alpha\, h_m|^n)^m},$$

(4.10)

$$K(h_m) = K_s \frac{((1 + |\alpha\, h_m|^n)^m - |\alpha\, h_m|^{n-1})^2}{(1 + |\alpha\, h_m|^n)^{m(l+2)}},$$

(4.11)

in which:
$\theta$  : moisture content (-);
$\theta_r$  : residual moisture content (-);
$\theta_s$  : saturation moisture content (-);
$\alpha$  : parameter ($cm^{-1}$);
$m$  : help parameter: $m = 1 - 1/n$ (-);
$n$  : parameter (-);
$l$  : parameter (-);
$h_m$  : matric pressure (cm);
$K$  : unsaturated hydraulic conductivity (cm/d);
$K_s$  : saturated hydraulic conductivity (cm/d).

In-depth studies have been carried out to determine Van Genuchten's parameters accurately (e.g. Wörsten et al., 1994).

### Percolation and Capillary Rise

Field capacity is the equilibrium situation in which the maximum quantity of water is contained in the unsaturated zone by capillary forces, notwithstanding the force of gravity. The field-capacity situation is rather unstable since it reflects a stationary situation, without downward or upward flow, which in practice hardly ever occurs. The amount of water stored in the soil at field capacity also depends on the actual groundwater depth. For these reasons the term should only be used to indicate a temporary situation as a starting point to describe subsequent processes.

If infiltration into the subsurface occurs and the unsaturated zone is not yet at field capacity, the soil will absorb all infiltrated water. Then, a new equilibrium state develops in which the water in the unsaturated zone is distributed over the entire height of the zone in accordance with the θ-$h_m$ relationship that is described by the retention curve. Once the unsaturated zone has absorbed as much water as possible, the situation of field capacity is reached. If more water infiltrates into the soil, it will flow down to the groundwater, consequently raising the groundwater table.

If water is abstracted from a soil that is in the situation of field capacity, the total moisture content decreases, the soil moisture following the retention curve. The positive potential difference in an upward direction is the driving force of capillary rise. As a result of this capillary rise, the groundwater table generally falls.

According to Darcy's law (Eq. 4.6), capillary rise depends both on hydraulic conductivity and matric pressure differences. In the dynamic situation, the capillary rise in the unsaturated zone is generally not the same over the entire zone height. During droughts, for instance, more water is abstracted from the root zone than can be replenished by groundwater.

## *Evapotranspiration*

Evapotranspiration is defined as the transpiration from plant leafs together with the evaporation from interceptions and bare soil surfaces. To determine the actual evapotranspiration, various variables have to be taken into account, such as:
- crop type;
- crop growing stage;
- plant cover of the soil;
- moisture content of the soil.

The evapotranspiration is given by:

$$E_t = E_p + E_i + E_s,$$
(4.12)

in which:

$E_t$ : evapotranspiration (mm/d);
$E_p$ : evaporation from the soil (mm/d);
$E_i$ : evaporation from interceptions (mm/d);
$E_s$ : transpiration from plants (mm/d).

In practice, evapotranspiration of plants and soil are derived from crop coefficients (Feddes, 1987). Crop coefficients can be determined by lysimeter tests, in which ideal growing circumstances are maintained such as optimal soil moisture content and groundwater depth.

Figure 4.7 shows the basal 'crop curve' during the various growing stages in one season, assuming dry soil conditions and an ideal amount of moisture available in the soil. In general, wet soil conditions result from precipitation or irrigation, which affect the total

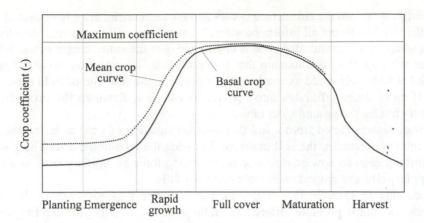

*Fig. 4.7. Crop curve as a function of growing stage (after Wright, 1982).*

evapotranspiration, especially when vegetation coverage is sparse. To incorporate the moisture content of the soil in the crop coefficients, a mean crop curve is being used to determine the evapotranspiration during the crop growing stages: the dashed line in Fig. 4.7.

Crop coefficients are available for various crop types, including the growing stages at fixed moments in the growing season. They reflect the mean curves, at wet soil conditions and average growing conditions. The use of crop factors enables determination of the potential evapotranspiration:

$$E_{tp} = Cf_m E_r, \qquad\qquad (4.13)$$

in which:

$E_{tp}$ : potential evapotranspiration (mm/d);
$E_r$ : reference evaporation (mm/d);
$Cf_m$ : mean crop coefficient (-).

Corrections for less ideal situations than the ones in lysimeters are included separately in the evapotranspiration calculation. Very important are the availability of water in the root zone and the capability of roots to abstract water from the root zone.

According to Feddes et al. (1988) less ideal situations with respect to the pressure in the root zone can be included empirically in the plant-transpiration calculation by the sink term variable α (Fig. 4.8).

As shown in Fig. 4.8, as a result of oxygen deficiency in the root zone, water uptake below $h_{mr} = h_1$ is zero. Above $h_4$, which represents 'wilting point', water uptake is also zero. The location of pressure $h_3$ is called 'reduction point'. Between $h_2$ and $h_3$, water uptake is maximal. The curve between $h_3$ and $h_4$ is considered to be linear and depends on potential plant transpiration.

Assuming that the uptake of water by the roots is equal to the transpiration and the root uptake is distributed equally over the root zone height, the following equations can be derived:

$$E_p \;=\; \alpha(h_{mr})\; E_{pp} \;=\; \alpha(h_{mr})\; Cf_b\; E_r, \tag{4.14}$$

$$S(h_{mr}) \;=\; \frac{E_p}{|z_r|}, \tag{4.15}$$

in which:
$E_p$ : transpiration from plants (mm/d);
$E_{pp}$ : potential transpiration from plants (mm/d);
$E_r$ : reference evaporation (mm/d);
$\alpha$ : sink-term variable (-);
$z_r$ : depth of the root zone (mm);
$S$ : sink term (d$^{-1}$);
$h_{mr}$ : average matric pressure in the root zone (cm);
$Cf_b$ : basal crop coefficient (-).

Equation 4.14 uses the basal crop coefficient, which is the coefficient described by the basal crop curve of Fig. 4.7. Beware of the difference between the basal crop coefficient and the mean crop coefficient. As shown in Fig. 4.7, the difference between the two coefficients can be considerable, especially when the soil is only sparsely covered.

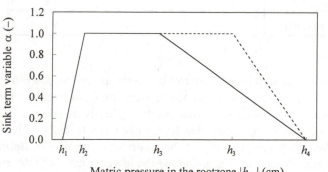

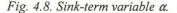

Matric pressure in the rootzone $|h_{mr}|$ (cm)

*Fig. 4.8. Sink-term variable $\alpha$.*

## 4.3  Flow Elements of a Water System

### 4.3.1  Types of Elements

As mentioned in the introduction to this chapter the following types of flow elements are distinguished in a water system:
- regulating structures,
- fixed structures,
- free flow elements.

The flow elements distinguished in this thesis are listed in Table 4.1. Of special interest in the table are the elements which can be regulated: pumping stations, weirs, inlets and sluices. Weir and inlet flow elements can be either be regulated or fixed. The majority of weirs and inlets in the Netherlands are of the fixed type. This means that they are manually adjustable, but that adjustment occurs very rarely, e.g. only twice a year at the transition from the summer to the winter season and vice versa.

Table 4.1. Flow elements distinguished.

|  | Regulating structure | Fixed structure | Free flow element |
|---|---|---|---|
| Pumping Station | Yes | | |
| Weir | Yes | Yes | |
| Sluice | Yes | | |
| Inlet / Outlet | Yes | Yes | |
| Canal | | | Yes |
| Groundwater | | | Yes |

All regulating structures can, in principle, control upstream as well as downstream surface-water levels. In general, pumping stations in polder areas are downstream-controlled since they control the lower water levels and discharge to higher-lying waters. In hilly areas, pumping stations are often upstream-controlled, to regulate water levels in higher-lying areas. If weirs are controlled, they are generally of the upstream-control type. Spilling sluices, which are mostly located along the coast and along the lower reaches of tidal rivers, are all of the upstream control type. Inlets are generally downstream controlled, whereas outlets are upstream controlled.

The capacity of regulating structures is always limited. During specific conditions, the capacity of structures is fully used in trying to maintain the required conditions in a controlled subsystem. The subsystem load is considered *high* in that situation.

### 4.3.2 Pumping Stations

Pumping stations are the main regulating structures in polder areas. The capacity of these structures ranges from below 1 m³/s for small pumping stations up to 50 m³/s for the large ones.

In general, large pumping stations have several pumping units, e.g. four. A pump may get out of order, for instance, if a mechanical or electrical part breaks down. To maintain the required discharge capacity, smaller pumping stations sometimes also have more than one pumping unit. Pumping stations can be either manually or automatically controlled.

In case of manual control, the status of the pumping station is entirely determined by the operators, who switch the electrical or diesel engines on or off. Under normal conditions, without high system loads, these pumping stations are operated only during the day-time on working days. The general operational strategy followed before weekends, is to pump out water in advance to give operators some days off. The limited hours available for manual control, generally result in a simple way of operation in which the maximum capacity of an entire pumping station is used frequently.

Pumping stations which are controlled automatically can be of the electrical or diesel-engine-driven type. They are in majority controlled locally and operated on the basis of water-level measurements in the direct vicinity of the structure. Automation is often associated with unattended operation, which means that in addition to the control equipment, warning systems with pagers are installed. The advantage of such systems is that operators can be somewhere else and still be warned when an alarming situation occurs at a pumping station.

Automatic control generally means continuously available pumping capacity and therefore, a more stable surface-water level (e.g. Fig. 1.10, Page 16).

Below, several important subjects of controlled pumping stations are discussed. Automatically controlled pumping stations are assumed.

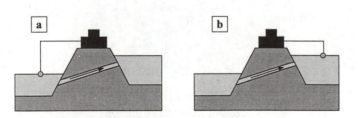

*Fig. 4.9. Upstream-controlled (a) and downstream-controlled (b) pumping station.*

In general, a locally controlled pumping station operates on the basis of only the upstream or the downstream water level. Figure 4.9 shows these two types of control. In both types, the operation of pumps is based on the actual water level measured. Figure 4.10 shows how switching takes place in a pumping station with three pumping units, in case of downstream control. Each of the pumps operates on the basis of its own on- and off-levels. As soon as the water level has reached its on-level for unit 1, the unit is switched on. If the pumping capacity is sufficient, the water level falls. Once the off-level is reached, the unit is switched off. If the capacity of a unit is not sufficient, a next unit is switched on when its on-level is reached. This process continues as long as more units are available.

In practice, restrictions are incorporated in the control of pumping stations. Pumping stations that drain polders and discharge to storage basins generally have extra conditions governing their operation. One restriction is the upstream water level in the storage basin. If this water level has reached an upper limit and is still rising, a milling stop can be imposed by the water manager of the storage basin. As a consequence, pumping stations should be switched off.

In case the pumps of a pumping station are driven by electrical energy, special measures may be taken to minimize energy costs. In general, the night tariff for electricity is lower than the day tariff. In that case, automated pumps should preferably operate during the night. To accomplish this, lower switching-on and -off levels can be set for the night.

Some electricity companies apply additional high tariffs for energy use during peak hours. In general, pumps are switched off during these hours. Under extreme conditions, the water manager may still decide to use these hours for pumping.

Frequent switching on and off of pumps is generally not allowed because of the chance of electrical overload and mechanical wear and tear. Especially for automated pumping stations, this is a matter of concern and special measures are taken to prevent continuous switching.

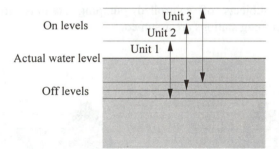

*Fig. 4.10. Control ranges and switching levels for upstream control using three pumps.*

### 4.3.3  Weirs

Weirs are used in water management in both polder and hilly areas, but they are mainly found in hilly areas. A weir can either have a fixed crest level or an automatically controlled crest level (Fig. 4.11). In the latter case, the water level upstream or downstream is controlled by means of a mechanical or electrical unit.

In general, electrical controllers are used these days for weir regulation. This requires on-line water-level measuring. The measured signal is fed to the controller, which determines the control action of the weir. The control action is electronically sent to the driving device of the weir. This device can adjust the weir in an upward or downward direction.

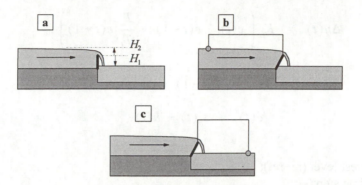

Fig. 4.11. Fixed (a), upstream-controlled (b) and downstream-controlled (c) weirs.

Du Buat (1816) determined a general weir equation for both submerged and unsubmerged weirs, which was adapted and evaluated by Franke (1970):

$$Q_w = C_w \, w_w \, (\Delta H + \frac{3}{2} H_1) \, \Delta H^{\frac{1}{2}}. \tag{4.16}$$

in which:
$Q_w$    : discharge over the weir (m³/s);
$C_w$    : discharge coefficient, ranging from 0.60 to 0.65 for sharp crested weirs (m^{½}/s);
$w_w$    : width of the weir (m);
$\Delta H$    : water-level difference (m);
$H_1$    : energy level to the crest downstream of the weir (m);
$H_2$    : energy level to the crest upstream of the weir (m).

The following two situations can be distinguished:
• if the downstream water level is below the weir crest, then $\Delta H = H_2$ and $H_1 = 0$;
• if the downstream water level is above the weir crest, then $\Delta H = H_2 - H_1$.

Devices used for automatically controlled weirs generally include controllers of the Proportional-Integral-Differential type (PID). Only the 'PI' actions of this controller are discussed here, since the 'D' action in water-system control may produce undesirable results, as will be discussed later.

The controller tries to establish a constant up- or downstream water level $h$ which meets a predefined setpoint $s_h$. The PI controller accomplishes this by a proportional offset, which represents the change in crest level on the basis of the difference (error) between the actual water level and water-level setpoint. Furthermore, it integrates that difference over time.

According to Åström & Wittenmark (1984) the output (here crest-level change) of a digital PI controller can be represented by the following equations:

$$\Delta u(t) = K_p \left( e(t) - e(t-1) + \frac{T_s}{T_i} e(t-1) \right), \tag{4.17}$$

$$u(t) = u(t-1) + \Delta u(t), \tag{4.18}$$

$$e(t) = s_h(t) - h(t), \tag{4.19}$$

in which:

| | | |
|---|---|---|
| $u$ | : | crest level (m+ref); |
| $t$ | : | time step (-); |
| $K_p$ | : | proportional gain (-); |
| $e$ | : | error (m); |
| $T_s$ | : | sample period (s); |
| $T_i$ | : | integral action time (s); |
| $s_h$ | : | water-level setpoint (m+ref); |
| $h$ | : | actual water level (m+ref). |

Equation 4.17 is known as the 'velocity form' of the digital PI controller. It determines the change in crest level rather than the crest level itself. The control range of a weir is generally limited by an upper and a lower crest level. In the velocity form of the controller, the restrictions to the crest-level movement can be incorporated by setting $\Delta u(t) = 0$, at the upper and lower limits.

The two parameters that determine operation of the controller are: the proportional gain $K_p$ and the integral action time $T_i$. These parameters should be determined carefully, since they determine the performance of the controller, e.g.: reaction speed and stability. The controller parameters depend on several properties of the controlled water system, e.g. the dimensions of the water course in which the weir is located. This subject has been studied by Schuurmans (1997). Translation waves resulting from weir adjustments may reflect at specific locations in a controlled canal and return to the weir, where they are detected by the water-level measuring equipment. Translation waves in a canal can thus result in system oscillations. The use of the differential 'D' action of the controller seems to enlarge the risk of oscillation, which is a reason why it is generally not used.

### 4.3.4 Sluices

Most spill sluices along the coast and tidal rivers are of the sliding-gate type in the Netherlands. The sluice structure consists of one or more culverts, each of which can be closed off by a sliding gate. The discharge of a sluice can be described by various water-level situations on both the upstream and downstream sides. The entire discharge process is rather complex and not easily described in one equation. In general, the main situations that can be distinguished to describe the flow via a sluice structure are (Fig. 4.12):

a.    unsubmerged underflow and submerged gate outflow;
b.    submerged underflow and submerged gate outflow;
c.    free gate underflow and unsubmerged weir overflow;
d.    free gate underflow and submerged weir overflow.

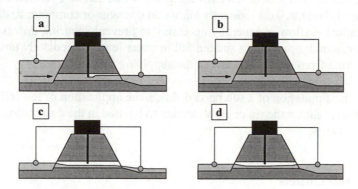

*Fig. 4.12. Main flow situations distinguished in a spill sluice.*

In situations (a) and (c) the discharge is only determined by the upstream water level, whereas in situations (b) and (d) it is determined by both the upstream and downstream water levels.

It is not the purpose of this thesis to describe the discharge through a sluice in detail. This has been done by various other authors (e.g. Spaan, 1994). For the present study, it is assumed that the sluice has an entirely submerged outflow, corresponding to situation (b) in Fig. 4.12. In that situation, the discharge can be determined by the orifice equation. Neglecting the effects of the number of gate openings, which contribute to the flow and the change in contraction in different operational situations, the general equation for flow through the sluice yields:

$$Q_{sl} = \mu_{sl} A_{sl} \sqrt{2g} \, \Delta H^{\frac{1}{2}},$$  (4.20)

in which:

| | | |
|---|---|---|
| $Q_{sl}$ | : | discharge via the sluice (m³/s); |
| $\mu_{sl}$ | : | contraction coefficient (-); |
| $A_{sl}$ | : | total cross-sectional area of sluice gates (m²); |
| $g$ | : | gravitational constant (m/s²); |
| $\Delta H$ | : | positive water level difference over the sluice (m). |

A controlled sluice can be opened as soon as the upstream surface-water level is higher than the downstream water level. Two control situations can be distinguished:

- the sluice is always opened in case of a positive water-level difference;
- the sluice is opened when a positive water-level difference occurs and, moreover, the water level inside the upstream polder is or will become too high.

An automatically operated sluice requires water-level measuring on both sides of the structure. In general, spill sluices have several gates which can be operated separately and be fixed at various heights. This indirectly allows an operator or computer to determine the discharge. Gradual outflow is generally necessary to prevent scouring downstream of the structure. It furthermore prevents a sudden fall in water level immediately upstream of the structure, resulting from inertia of water in the supplying water course.

In addition to the simulation of a submerged sluice, the application of the orifice equation as described above, allows the sluice flow element to be used in the description of restricted flow through culverts.


### 4.3.5  Inlets and Outlets

Inlet and outlet structures function very similar to sluices. Inlets and outlets are used for water-quantity and water-quality control. The function of inlets and outlets is very similar and they are therefore described together in this thesis, using the term 'inlets' only.

Inlets are generally culvert-type structures that can control the downstream or upstream surface-water level (Fig. 4.13). Inlets discharge by gravity. In this thesis the inlet flow element is used to describe fixed in- or outflow. If the flow via the inlet depends on the water levels upstream and possibly also downstream, the sluice flow element should be used.

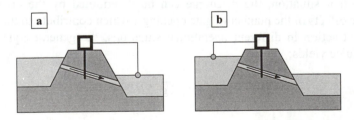

*Fig. 4.13. Downstream-controlled (a) and upstream-controlled (b) inlets.*

### 4.3.6 Canals

The surface-water flow in regional water systems can be represented by the one-dimensional flow equations of Saint Venant. These equations comprise conservation of mass and mass-momentum descriptions, which can be solved numerically. Various forms of the equations can be found in literature, using several simplifying assumptions (e.g. Verspuy & De Vries, 1981). The equations are also known as the dynamic wave equations and for one-dimensional flow through an open prismatic canal can be written as:

$$\frac{\partial Q}{\partial x} + \frac{\partial A}{\partial t} = 0, \qquad (4.21)$$

$$\frac{\partial Q}{\partial t} + \frac{\partial}{\partial x}\left[\beta \frac{Q^2}{A}\right] + gA\frac{\partial h}{\partial x} - gAI_b + \frac{gQ|Q|}{C^2AR} = 0, \qquad (4.22)$$

in which:

| | | |
|---|---|---|
| $Q$ | : | flow (m³/s); |
| $t$ | : | time (s); |
| $x$ | : | axis in the direction of flow (m); |
| $\beta$ | : | Boussinesq coefficient (-); |
| $A$ | : | cross-sectional area (m²); |
| $g$ | : | gravitational constant (m/s²); |
| $h$ | : | water depth (m); |
| $I_b$ | : | bed slope (-); |
| $R$ | : | hydraulic radius (m); |
| $C$ | : | Chézy coefficient (m$^{½}$/s), $C = 18\,^{10}\!\log(12\,R/k_N)$; |
| $k_N$ | : | Nikuradse roughness coefficient (m). |

In Eq. 4.22, the first two terms are called inertia terms and represent the influence of acceleration and velocity head; the third and fourth term are called pressure terms, which incorporate the influence of the surface-water gradient and gravity forces; the fifth term, called the friction term, represents the change in flow as a result of friction losses.

Several numerical models have been built to solve Eq. 4.22 entirely or partly. In the latter case, inertia terms and/or pressure terms are neglected. It is not the aim of the present study to replicate that work, but to demonstrate that operational strategies can be formulated, including free flow elements. A first-order approach is chosen and therefore, in the mathematical model built in this thesis, only the last two terms are used, representing the equation of Chézy.

### 4.3.7 Groundwater

This section focuses on regional groundwater flow which occurs over long distances, such as between polders or between water systems. Two basic types of groundwater flow are distinguished here: unconfined flow through phreatic aquifers and confined flow within aquifers between impermeable layers in the subsurface. Seepage from and infiltration into semi-confined aquifers are considered constant during the control periods of interest (hours to weeks). Local groundwater flow has been described in Sec. 4.2.2.

Groundwater flow is determined by differences in water potential between parts of the soil or adjacent surface water. According to Darcy's law, the saturated flow through a soil can be described by:

$$
\begin{aligned}
q_x &= -K_x \frac{\partial h}{\partial x}, \\
q_y &= -K_y \frac{\partial h}{\partial y}, \\
q_z &= -K_z \frac{\partial h}{\partial z},
\end{aligned}
\tag{4.23}
$$

in which:
$q$      : flow (m/s);
$x$      : $x$-direction (m);
$y$      : $y$-direction (m);
$z$      : $z$-direction (m);
$K$      : hydraulic conductivity (m/s);
$h$      : water potential (m+ref).

When no water is abstracted from, or recharged to the groundwater system, the conservation of mass yields:

$$
\frac{\partial q_x}{\partial x} + \frac{\partial q_y}{\partial y} + \frac{\partial q_z}{\partial z} = 0.
\tag{4.24}
$$

The Darcy and mass equations can be combined into the Laplace equation:

$$
\frac{\partial^2 h}{\partial x^2} + \frac{\partial^2 h}{\partial y^2} + \frac{\partial^2 h}{\partial z^2} = 0.
\tag{4.25}
$$

In the present study, groundwater flow is described in one direction at a time and in that case the solution to the Laplace equation is linear in $s$ ($x$, $y$, or $z$) and yields:

$$
\begin{aligned}
h &= h_1 + (h_2 - h_1)\frac{x}{L}, \\
q &= -K_s \frac{h_2 - h_1}{L}
\end{aligned}
\tag{4.26}
$$

in which:

$h_1$       : water potential in location 1 (m+ref);
$h_2$       : water potential in location 2 (m+ref);
$s$       : direction of flow ($x$, $y$ or $z$) (m);
$L$       : distance between locations 1 and 2 (m).

The situation described with groundwater flow in one direction can be compared with the flow through a series of uniform tubes, together called an arc. By means of a network of arcs, regional groundwater flow can be described in a general and very flexible way. Similarly both horizontal flow through aquifers and vertical flow through aquitards can be incorporated in the modeling. The method can be further generalized, using soil-resistance terms as applied in the soil in- and outflow equation of Eq. 4.1. However, this approach has not been attempted in the present study.

## 4.4   Concluding Remarks

The basic equations that describe flow processes in water systems and the equations used in this thesis for modeling have been given in this chapter. It should be stressed that the approach presented enables incorporation of more or fewer modeling elements, depending on the required detail. For analysis purposes, however, it is most convenient to restrict the size of models and generalize the water system layout into a limited number of subsystems and flow elements. This will be shown in Chapter 7.

The computer program AQUARIUS has been developed especially for water-system modeling and analysis. The program dynamically calculates the behavior of all elements of a water system, the water-system load and the required operations of the regulating structures. AQUARIUS enables deterministic water-system modeling and time-series simulation and is based on the theory outlined in this chapter and furthermore on the mathematical equations given in Chapters 6.

# 5 Hydrological Load

## 5.1 Introduction

This chapter describes the hydrological load on water systems, which is important for both simulation on the basis of actual hydrological conditions and prediction of the system behavior on the basis of predicted hydrological conditions.

The first part of this chapter presents an analysis of the hydrological time-series data which were available to the present study. The purpose is to derive the necessary statistical information on hydrological loads. Furthermore, periods with excessive precipitation and evaporation are identified. These extremes are used to assess the performance of the control systems described in the practical case studies in Chapter 7.

The second part of this chapter focuses on the possible methods for predicting the hydrological load on a water system. The analysis and real-time situations are described separately, since they require a different approach in using a prediction of the hydrological load in the methodology developed.

## 5.2 Precipitation and Evaporation Analysis

The analysis of the hydrological load is based on 30-year time series of hourly precipitation and daily reference evaporation data measured at the De Bilt meteorological station. These time series were supplied by the Royal Netherlands Meteorological Institute (KNMI). In the following, the 30-year time series of De Bilt will be indicated by *De Bilt time series*.

The average yearly precipitation in the Netherlands is between 750 to 800 mm. It should be realized that precipitation characteristics show enormous differences from year to year. Figure 5.1 shows precipitation totals in the period of 1965 to 1994. The annual precipitation in this period ranges from 550 to 1150 mm. The years 1965 and 1966 were extremely wet, whereas the years 1971, 1976 and 1982 were extremely dry.

In this thesis, the reference evaporation defined by Makkink is used, for reasons explained below. The variation in reference evaporation is far less than that in precipitation. Figure 5.1 shows that the annual reference evaporation ranges from 500 to 625 mm. An interesting phenomenon is that years with high precipitation show a relatively low reference evaporation.

The following hydrological *seasons* are defined here and will be used in the remainder of this thesis:

- Winter:       1 January - 31 March;
- Spring:       1 April - 30 June;
- Summer:       1 July - 30 September;
- Autumn:       1 October - 31 December.

Spring and summer together are also indicated as the *summer season*, while autumn and winter together are also indicated as the *winter season*.

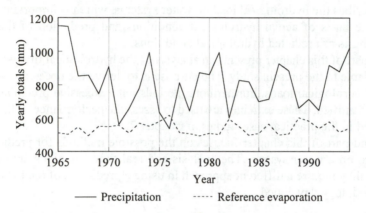

*Fig. 5.1. Yearly precipitation and reference evaporation (De Bilt time series).*

### 5.2.1 Precipitation

Precipitation mostly occurs as rain and to a lesser extent as snow, dew, condensation on ice, fog and frost. The clouds from which this precipitation falls, can be subdivided into 'stratiform' and 'convective' forms (Klaassen, 1989). Stratiform clouds have an horizontal shape and are generally layered. Precipitation from stratiform clouds can persist for long periods, e.g. 24 hours or longer (Wartema, 1989) and be wide spread, the rain falling in areas of several hundreds of kilometers.

Convective clouds have a vertical orientation and develop as a result of an unstable atmosphere in which warm air rises. The rise is generally a result of the difference in temperature between the warmed up lower layers of the atmosphere, close to the earth surface, and the relatively cold higher layers. The stronger the vertical air movement, the longer and more intense the convective precipitation can be. Convective precipitation generally falls over a limited area up to several kilometers and can be very local but of an extremely high intensity. The area of influence is generally limited to a size of several kilometers. Individual showers from convective clouds generally exist for short periods of an hour or even shorter (Van de Berg, 1989). However, new clouds may develop rapidly, and therefore convective precipitation may last for several hours.

Stratiform and convective precipitation often occur together. In a stratiform precipitation front, small convective areas with high-intensity precipitation may occur. In general, convective precipitation falls mainly in summer, whereas stratiform precipitation falls mainly in winter.

Figure 5.2 shows the 10, 25, 50, 75 and 90% percentile curves of precipitation over a year. To balance the effect of the number of days per month, in the overall picture, the unit used for precipitation is mm/day. Extreme events are found to be more pronounced during the months July to December.

Precipitation intensity on average slightly increases from winter to autumn. There is a distinct difference in intensity, which can be observed when looking at separate events. To study this, the 30-year time series of precipitation data is further analyzed. The following definitions with respect to precipitation events are used:
- a 'precipitation event' is a period of precipitation, which may include dry periods up to a maximum length of 2 hours;
- precipitation events with a total volume smaller than 0.3 mm are neglected.

This analysis yields an average precipitation of 464 hours a year, distributed over an average number of 197 events. On average, therefore it rains in the Netherlands 5.3% of the time.

Figure 5.3 shows that the precipitation intensity per event increases during spring, showing the highest intensities in the summer, after which it decreases again in autumn. Usually, precipitation in summer is of a short duration. The overall picture, therefore, shows short and heavy summer showers, resulting from convective clouds.

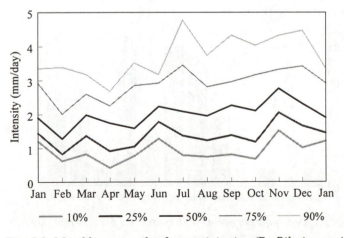

*Fig. 5.2. Monthly percentiles for precipitation (De Bilt time series).*

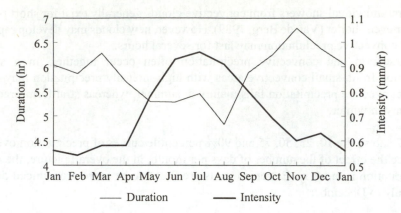

*Fig. 5.3. Average duration and intensity of precipitation events (De Bilt time series).*

To find the total precipitation quantity per event, the extremes of the De Bilt time series were further analyzed. Figure 5.4 shows 43 extreme precipitation events that resulted in quantities over 30 mm over the 30 years. The figure clearly shows that most extreme precipitation events happen at the end of spring and in summer.

To compare the results of various modes of control, several shorter periods from the entire time series were studied in detail. Table 5.1 presents the most extreme events.

Two interesting events should be mentioned: the first in 1965, when 56.5 mm fell in winter over more than two days; the second in the summer of 1966, when 48.1 mm fell in a period of only five hours.

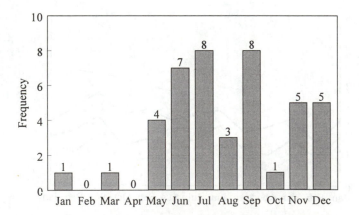

*Fig. 5.4. Precipitation events with quantities over 30 mm (De Bilt time series).*

Table 5.1. Extreme precipitation events with quantities over
45 mm (De Bilt time series).

| Start of event | Precipitation (mm) | Duration of event (hr) |
|---|---|---|
| 16-06-65 18:00 | 528 | 16 |
| 29-11-65 08:00 | 483 | 43 |
| 17-12-65 10:00 | 565 | 54 |
| 19-07-66 18:00 | 481 | 5 |
| 10-03-81 04:00 | 476 | 49 |
| 26-11-83 20:00 | 518 | 21 |
| 26-12-94 23:00 | 500 | 40 |

## 5.2.2 Evaporation

All over the world, various evaporation equations are used. The one most commonly used
is the Penman equation (Penman, 1948) or derivatives of this equation:

$$E_0 = \frac{\Delta}{\Delta + \gamma} \frac{(1 - \alpha) R_s - R_{nl}}{\lambda} + \frac{\gamma}{\Delta + \gamma} E_a,$$
(5.1)

in which:
$E_0$     : evaporation from an open water surface (kg.m$^{-2}$.s$^{-1}$);
$R_{nl}$    : outgoing long-wave radiation (W/m$^2$);
$R_s$     : incoming solar radiation (W/m$^2$);
$\alpha$      : Albedo reflection coefficient (-);
$E_a$     : aerodynamic evaporation equivalent (kg.m$^{-2}$.s$^{-1}$);
$\Delta$      : slope of the vapor pressure curve (kPa/°C);
$\gamma$      : psychometric constant (kPa/°C);
$\lambda$      : latent heat of vaporization (J/kg).

In the present study the Makkink equation is used, which determines the reference
evaporation $E_r$. The original equation of Makkink (1957) reads:

$$E_r = C_1 \frac{\Delta}{\Delta + \gamma} \frac{K^{\downarrow}}{\lambda} + C_2,$$
(5.2)

in which:
$E_r$     : reference evaporation (kg.m$^{-2}$.s$^{-1}$);
$K^{\downarrow}$     : global radiation density (W/m$^2$);
$C_1$     : parameter (-);

$C_2$      : parameter (-);
$\Delta$      : slope of the vapor pressure curve (kPa/°C);
$\gamma$      : psychometric constant (kPa/°C);
$\lambda$      : latent heat of vaporization (J/kg).

The reason for using the Makkink evaporation equation is that all weather bureaus in the Netherlands and the end users of these data, e.g. farmers, use it too. In practice, these are the only data available to base operational control strategies on.

The KNMI weather bureau gave the following reason for switching from the Penman evaporation to the Makkink evaporation in 1987 (KNMI/CHO, 1988):
• the calculations involved in the Penman equation are complex: various parameters of the equation are hard to determine;
• a single, standardized, evaporation equation should be used, instead of the various versions of the Penman equation, in use all over the world.

The Makkink equation basically requires only two variables: global radiation and air temperature. These variables can be measured with relative ease. In 1987 the KNMI introduced the following form of the Makkink equation, which contains only one dimensionless parameter $C$:

$$E_r = C \frac{\Delta}{\Delta + \gamma} \frac{K^{\downarrow}}{\lambda}.$$

(5.3)

The evaporation calculated by Eq. 5.3 is in approximately 1.17 to 1.31 times as high in spring and summer than the evaporation calculated by the Penman equation. For open water the ratio Makkink evaporation / Penman evaporation is approximately 1.25. This difference has been incorporated in the crop coefficients used with the Makkink evaporation equation during the growing season (Feddes, 1987).

Figure 5.5 shows an example of a typical evaporation pattern over a year. The influence of temperature can be seen clearly: low evaporation rates in winter and high evaporation rates in summer.

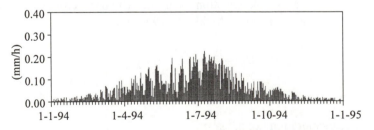

Fig. 5.5. Daily reference evaporation in 1994 (De Bilt time series).

High evaporation especially affects hilly areas that suffer from a water shortage in summer. Analysis of the De Bilt time series shows several periods of extremely high evaporation. For 55 days a reference evaporation exceeding 5 mm has been found. Table 5.2 shows the highest extremes in evaporation.

A remarkable period of extreme evaporation occurred during the summer of 1976 when, for a consecutive period of 11 days, the evaporation exceeded 5 mm/day. As can be seen in Fig. 5.1, 1976 indeed was a very dry year; the only year when the reference evaporation exceeded the annual precipitation.

Table 5.2. Extremes in reference evaporation
(De Bilt time series).

| Start of event | Maximum reference evaporation (mm/day) | Number of consecutive days |
|---|---|---|
| 29-06-76 | 5.7 | 11 |
| 10-07-82 | 5.5 | 1 |
| 28-06-86 | 5.4 | 4 |
| 17-06-89 | 5.4 | 1 |
| 20-06-89 | 5.4 | 1 |
| 34674 | 5.4 | 2 |

## 5.3    Prediction of the Hydrological Load

In the control of water systems that have rapid runoff characteristics, a hydrological-load prediction may be necessary for part of the control horizon. In that case, a forecast of hydrological variables is required.

Since the prediction of the hydrological load is an estimate which involves several uncertainties, it has to be updated each time additional information about the current hydrological variables becomes available, e.g. precipitation per hour and evaporation per day. In real-time control, these variables are forecasted by meteorologists of a weather bureau. However, for analysis purposes, historical data from a database are used for prediction here.

In this thesis, a distinction is made between the 'prediction' of hydrological variables and the 'forecast' of these variables. Prediction is used as general term for all kinds of determinations of hydrological variables for the control horizon. The word forecast is used exclusively for the hydrological variables provided by a weather bureau.

A weather forecast is generally compiled by meteorologists on the basis of measured numerical data, graphical data (e.g. satellite and radar images) and the outcome of numerical atmosphere models. Monitoring data used by weather bureaus are obtained from world-wide monitoring networks. Such networks consist of monitoring stations all over the earth and in the air and include radar units and satellites. Important variables monitored are: temperature, air pressure, wind velocity, wind direction, and air humidity.

In the following, the subjects of weather forecasts for analysis purposes and real-time control are discussed separately, since they require principally different approaches. Weather-forecast data are described only qualitatively, to give an impression of how forecasted data can be incorporated in real-time control. A theory is developed that allows basically inaccurate forecasts to still be incorporated into real-time control, with the lowest risk of undesirable results in control.

Table 5.3 lists the combinations of processing methods which are applied for the analysis application of the method developed and the equivalent data sources of forecasts in real-time control.

Table 5.3. Data sources and processing methods for hydrological-load prediction.

| Method | Application | |
|---|---|---|
| | Analysis | Real-time control |
| Perfect prediction (one time step) | Database (uncertainty multiplier) | Weather bureau (risk multiplier) |
| Period-average prediction (several time steps) | Database (uncertainty multiplier) | Weather bureau (risk multiplier) |
| Scenario prediction | Time-series analysis | Time-series analysis |

## 5.3.1  Prediction for Water-System Analysis

### Methods

In the water-system analysis presented in this thesis, hydrological variables are predicted by the prediction module of the Strategy Resolver on the basis of hydrological data from a database (see Fig. 3.3, Page 62). The data from this database are used to simulate weather forecasts by a weather bureau. For the time-series calculations, these data are retrieved from the database in advance, simulating forecasts. The simulated forecasts are processed to hydrological-load predictions for the control horizon, using the methods listed in Table 5.3.

The 'perfect prediction' simulates forecasts and assumes the required hydrological variables to be available for every time step of the control horizon, e.g. for every hour.

The 'average prediction' permits simulation of a forecast for a period which is longer than the optimization time-step size, e.g. the prediction of precipitation for every 6 hours, while the optimization step size is only one hour.

To simulate the uncertainty encountered in real weather forecasts, the *uncertainty multiplier* is introduced for perfect and period-average prediction methods. The uncertainty multiplier is used to manipulate and reduce the values obtained from the database to achieve a realistic prediction of the hydrological load for the control horizon in determining a control strategy. The uncertainty multiplier depends on the data obtained from the database and is generally smaller than unity. The prediction of precipitation for the control horizon can thus be determined by:

$$P_s(t) = u_m^* P_d(\tau + t),$$ (5.4)

$$u_m^* = f(P_d, T),$$ (5.5)

in which:

$P_s$     : precipitation used in strategy determination (mm);
$t$     : time step of the control horizon (-);
$u_m^*$     : uncertainty multiplier (-);
$P_d$     : precipitation obtained from database (mm);
$\tau$     : simulation time step (-);
$T$     : control horizon (-).

The 'scenario prediction' is especially suitable for periods for which generally no accurate predictions can be given, e.g. a period of a week. The scenario prediction is based on the statistical analysis of historical hydrological data. Hydrological conditions are classified into the categories: 'extremely wet', 'wet', 'normal', 'dry' and 'extremely dry' with respect to the time of year.

The graphs in Fig. 5.6 show examples of the basic data determined in the statistical analysis and used for scenario prediction. The curves have been smoothened. The graphs clearly show

Table 5.4. Scenarios for predicting the hydrological load.

| Scenario | Percentile values | |
|---|---|---|
| | Precipitation | Evaporation |
| Extremely wet | 90% | 10% |
| Wet | 80% | 20% |
| Normal | 50% | 50% |
| Dry | 20% | 80% |
| Extremely dry | 10% | 90% |

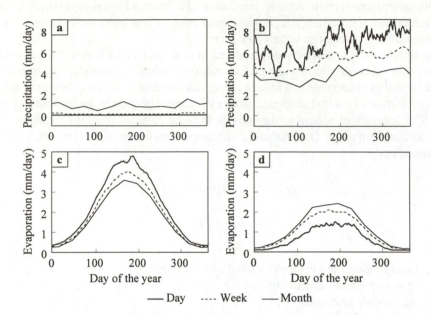

*Fig. 5.6. Percentile curves for periods of a day, a week and a month for precipitation:*
*10% (a) and 90% (b), and evaporation: 90% (c) and 10% (d) (De Bilt time series).*

that the shorter the prediction period (e.g. a day), the more extreme the predicted quantities of both precipitation and evaporation are (thick lines in graphs (b) and (c)). On the basis of similar percentile data, the scenarios of Table 5.4 were determined. In the application of these scenarios, it is assumed that high evaporation is associated with low precipitation and vice versa, a phenomenon, which has been explained in Sec. 5.2.

### *Application*

In the present study, a combination of the prediction methods presented in Table 5.3 is used. Figure 5.7 shows an example of how a prediction of precipitation could look, using the three prediction methods: perfect, period-average and scenario. A provisional uncertainty range is indicated in the figure. Precipitation used in strategy determination is chosen from that range.

In the example, during the first day, a perfect prediction for six-hour intervals is available. The perfect and period-average predictions can be used in combination with the uncertainty multiplier. A time-series example of this type of prediction is presented in Sec. 7.1.3. For the second and third period, each covering one day, a period-average precipitation prediction is shown. The fourth period shows the result of a scenario prediction.

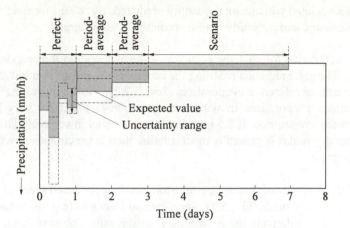

*Fig. 5.7. Precipitation prediction using three different methods:*
*perfect, period-average and scenario.*

### 5.3.2  Weather Forecasts for Real-Time Control

*Accuracy Considerations*

The accuracy of the precipitation and evaporation forecast determined by a weather bureau
for temperate zones depends on several factors, some of which are:

- the type of weather;
- the time of year;
- the size of the area for which a prediction should be made;
- the intensity and quantity predicted;
- the period for which a prediction should be made.

As discussed in Sec. 5.2, a distinction should be made between wide-spread stratiform
precipitation with low intensities and convective local precipitation in showers. Stratiform
precipitation from cloud fronts occurs mainly in winter. Convective precipitation from
compact and vertically oriented clouds occurs mainly in summer. Convective precipitation
can be very local, showing high intensities at some locations in the water system, while at
other locations no precipitation may occur at all. The two types of precipitation can occur
simultaneously.

Precipitation can vary strongly during the day and the exact moment of precipitation
is difficult to predict for long periods ahead, especially in summer. The probability of
precipitation falling in a specific area, depends on the size of the area considered. During
summer, local differences in precipitation prevail. High-intensity events are more difficult
to predict accurately than low-intensity events (De Rooy & Engeldal, 1992). This

phenomenon is associated with the total quantity predicted, e.g. a rainstorm during which 40 mm falls in four hours, can generally not be predicted at all accurately.

Similar to the precipitation forecast, the accuracy of the evaporation forecast depends on the type of weather. The solar radiation reaching the earth surface and air humidity, are the main factors that determine reference evaporation (Sec. 5.2.2). Furthermore, the time of year strongly determines evaporation. In winter, evaporation is generally very low to zero, whereas a reference evaporation of 2.5 mm/day is not unusual in summer. The variation in evaporation from day to day is generally much smaller than in precipitation, which makes it less difficult to predict.

On the basis of the accuracy of forecasts, a division into four preferred forecast methods and associated periods can be made (Fig. 5.8). Short-period forecasts (e.g. up to two hours) can be compiled with a relatively high accuracy using radar observations, possibly in combination with precipitation measurements (Van den Assem, 1989; Van der Beken, 1979).

Forecasts for periods up to one day ahead are most accurately deduced from numerical regional atmosphere models. Present-day models have a grid size of the order of magnitude of 50 × 50 km and use the results of world-wide atmospheric models as boundary conditions (Jilderda et al., 1995).

Forecasts for periods of several days ahead (e.g. one to five days) become the most accurate when world-wide atmospheric models are used (Hafkenscheid, 1988; Jilderda et al., 1995).

Forecasts for several days to weeks ahead are most accurately compiled by stochastical models (Buishand & Brandsma, 1996).

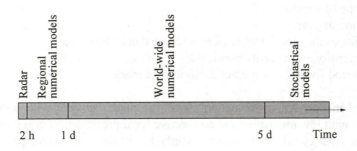

*Fig. 5.8. Forecast periods and most suitable methods for forecasting.*

Several studies have been carried out to verify the accuracy of forecasts that result from different forecast methods. In the literature found, only generalized forecast data have been used for the periods of interest to the present study.

Verification studies (e.g. Donker, 1989) generally find that meteorologists tend to underestimate precipitation in their forecasts, a phenomenon which increases with the quantity forecasted. This will be qualitatively demonstrated below. Daan (1993) published verification results of the KNMI, which show that the reliability of a precipitation forecast strongly depends on the time of year. He proves that the most inaccurate results are obtained for the summer season. This is probably due to the mainly convective character of precipitation during summer.

Figure 5.9 tentatively presents the reliability of forecasts as a function of the forecast period. Reliability is defined as the probability that the actual quantity of precipitation falls within the predicted range, e.g. 0 - 0.3 mm, 0.3 - 1 mm, 1 - 3 mm, 3 - 10 mm, > 10 mm. The general trend presented by the curve in the figure is that the reliability of a forecast for a very short period (hours) is relatively high, whereas the reliability decreases when the period increases to days. For very long periods (e.g. months), the reliability of the forecast increases again.

This phenomenon affects the determination of control strategies for various types of areas distinguished in this thesis. As will be shown in Chapter 7, for polder areas, a prediction period of zero up to several days is generally needed, whereas for hilly areas, predictions of several days to weeks ahead may be necessary.

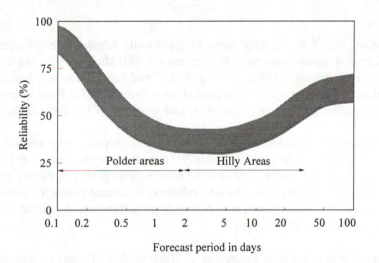

*Fig. 5.9. Reliability of the forecast of hydrological variables (indicative).*

### Risk-Based Approach

This section presents a theory on how inaccurate predictions can be incorporated in dynamic control, keeping the risk of undesirable situations the lowest possible.

Suppose a large quantity of precipitation is predicted to fall in a polder area and it is decided to pump out water in advance. By using the risk-based approach, the risk of damage to interests as a result of pumping out water, while the exact quantity of precipitation that will actually fall is not known, is minimized.

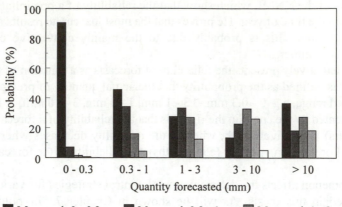

*Fig. 5.10. Precipitation forecasted and the probability of the forecast coming true (12 hour-forecast data set from De Rooy & Engeldal, 1992).*

The bar graph of Fig. 5.10 presents some of the results which were obtained in the verification study of a regional numerical forecast model. This study was carried out for the KNMI by De Rooy & Engeldal (1992) for the period of mid July 1991 to mid January 1992. The horizontal axis shows the quantity of precipitation forecasted for fixed intervals. The vertical axis shows the probability of the measured quantity within each of the forecast intervals.

The bars show that forecasts for 0 - 0.3 mm (no precipitation) have a relatively high probability of being correct. The right-hand bars (> 10 mm) show that none of the forecasts for > 10 mm came true. This proves that extreme precipitation events cannot be forecasted accurately. Moreover, when analyzing the total number of extreme events in the data set of half a year, the number of forecasts for > 10 mm is 40% less than the actual number of precipitation measurements of > 10 mm.

A weather forecast should basically be used in real-time control of water systems that have relatively rapid runoff characteristics. The forecasted hydrological variables for a period within the control horizon are obtained from a weather bureau. At present, weather bureaus can supply these data with a small resolution of fractions of one hour and update them several times a day.

Since a forecast may incorporate a rather large uncertainty, the risk of control actions based on incorrect forecasts should be assessed and minimized, before control actions are taken. Using a specific control mode, the risk of damage to the interests in a water system during the control horizon can be represented by:

$$R_D = \sum_{t=1}^{T} \left[ P(P_a(t) \mid P_f(t)) \sum_{j=1}^{n} D_j(t) \right], \tag{5.6}$$

in which:
$R_D$      :   risk of damage (-);
$t$       :   time step (-);
$T$      :   control horizon (-);
$P$      :   probability (-);
$P_a$     :   precipitation to account for (mm);
$P_f$      :   precipitation forecasted (mm);
$j$       :   interest counter;
$n$      :   number of interests in the water system;
$D$      :   damage to an interest (-).

In this equation, the damage to interests is used in the form that will be explained later in Sec. 6.2.3 and 6.2.4. In real-time control, the inaccuracy of a forecast can be incorporated in a control strategy via the 'risk multiplier', which is applied to the forecasted precipitation:

$$P_s(t) \;=\; r_m^* \, P_f(t), \tag{5.7}$$

$P_s$     :   precipitation used in strategy determination (mm);
$t$      :   time step (-);
$r_m^*$     :   optimal risk multiplier (-);
$P_f$     :   precipitation forecasted (mm).

The risk multiplier is a function of several variables, amongst which: the precipitation quantity forecasted, the system state, the potential damage to interests and the control mode used. The optimal risk multiplier depending on these factors can thus be expressed by:

$$r_m^* \;=\; f(P_f, \bar{v}, D, CM, T), \tag{5.8}$$

in which:
$\bar{v}$      :   vector of water-system variables;
$D$      :   damage to water-system interests (-);
$CM$   :   control mode;
$T$      :   control horizon (-).

Figure 5.11 indicatively presents the risk of damage to interests as a function of the risk multiplier. Note that a large amount of verification data is needed to accurately determine the shapes of the curves. The maximum risk has been set at 100 for each time step, since the maximum dimensionless damage to all interests together is also set at 100 for each time step, as will become clear in Sec. 6.2.4.

The dark curve in the graph shows that when a low quantity of precipitation has been forecasted, the risk of damage to the interests is lowest when no precipitation is accounted for in the determination of the control strategy (risk multiplier = 0). However, in forecasts of extreme events, the risk involved when underestimating precipitation can be so high, that a risk multiplier larger than unity should be used.

The example given in the figure should be considered indicative. The subject requires further study. It shows, however, that simply using the forecast of a weather bureau in determining of a control strategy, is generally not ideal and makes the risk of performing incorrect control actions higher than necessary.

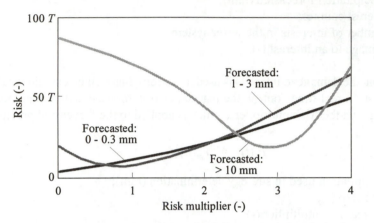

Fig. 5.11. Determining the risk multiplier associated with the lowest risk of damage to interests (indicative).

## 5.4    Concluding Remarks

Historical records show a great difference between hydrological conditions in the summer and winter seasons. In water-system analysis, extremes that occur infrequently are especially important, since these typically induce water-system failure. Several extreme situations which have occurred in the past, have been incorporated in the examples given throughout this thesis.

Various elements of the prediction module of the Strategy Resolver have been discussed in this chapter. For real-time control, weather forecasts have to be obtained from a weather bureau. In this thesis, the basis for the prediction module is a hydrological database. Both data obtained from a weather bureau and from a database have to be processed and incorporated into the hydrological load on the water system in order to obtain a reliable control strategy that minimizes damage to interests.

In general, the required prediction reliability should be determined for each specific water system and be based on a risk analysis. In such an analysis, the risk of incorrect predictions is assessed. A similar risk analysis has been carried out within the framework of the current study by Botterhuis (1997) for the Fleverwaard water system. It will not be discussed further in this thesis, however.

In optimization, stochastical distributions of precipitation and evaporation can be incorporated, using chance-constrained optimization. However, for the analysis purposes presented in this thesis, the expected values for precipitation and evaporation obtained are used for the perfect and period-average predictions in combination with the uncertainty multiplier. Furthermore, scenario predictions are used. The sensitivity to the accuracy of the predictions of the control strategies thus obtained, is assessed in the practical case studies of Chapter 7.

# 6 Optimization Problem

## 6.1 Introduction

In Chapter 2, the various tasks of a control system were discussed. The Strategy Resolver was introduced for solving an operational control problem for a certain period ahead. Chapter 3 gave a general description of the control problem and possible solutions. Successive Linear Programming was preferred as method. This mathematical optimization method enables large water-resources problems to be solved rapidly.

In this chapter, application of the SLP method is described and some enhancing features are discussed. Special points of interest in the discussion are the speed in solving the optimization problem and the ease of understanding how to build the optimization model. Both points are of great importance for the practical use of the method and for further research into this topic.

At the end of Chapter 3 the various modules of the Strategy Resolver, which are used to formulate and solve the control problem, were described in a general way. These modules and their interactions are shown in Fig. 6.1 in more detail. During operation, the models successively use data from each other. This process is called *simultaneous simulation and optimization*.

The simulation module determines the water-system state accurately. In practice this module could be replaced by a monitoring network, if all the required data would be available. The optimization module has the specific task to determine control strategies. The optimization problem involved consists of a generalized copy of the simulation model. In this study, the simulation only uses the set of control actions that belong to optimization time step $t = 1$, in accordance with the optimal control strategy.

The optimization module consists of two submodules called the *Constraint Manager* and the *Solver*. All preparations for solving the optimization problem are executed by the Constraint Manager. During the simulation and optimization loop, constraints may vary, e.g. the availability of regulating structures, updated linearized process descriptions. Furthermore, the presence and weights of objectives may vary in the course of time, because conditions change, e.g.: the growing season for agricultural interests; weekends and holidays for recreational interests. These and other constraints are updated by the Constraint Manager each loop.

This chapter focuses on the preparatory task of the Constraint Manager in formulating the optimization problem. The following objectives have been set for the Constraint Manager:

- the problem should be formulated in such a way that it can be solved rapidly;
- the problem should be kept as small as possible;
- the water-system process descriptions involved should be as accurate as possible.

The speed in solving an optimization problem depends on problem formulation and the type of solver used. The mathematical aspects of the solving method itself are outside the scope of the current study, and are discussed only briefly in Sections 6.2.8 and 7.5.

Keeping the optimization problem as small as possible means modeling only those processes which can be controlled directly or indirectly by means of regulating structures. When this is applied in the optimization model, a simplified description of the water system is all that is left, using the uncontrolled flows to and from controllable subsystems as boundary conditions.

Keeping the process descriptions in optimization accurate implies that the SLP method chosen requires much effort to achieve accurate linearization. This subject is described in Sections 6.3 and 6.4 in detail.

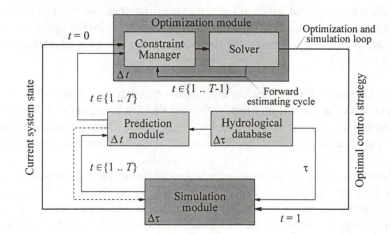

*Fig. 6.1. Interactions between the modules of the Strategy Resolver.*

## 6.2 Formulating the Linearized Problem

### 6.2.1 The Linear Model

In Sec. 3.3, the following general mathematical form of the optimization problem has been introduced:

$$\text{Minimize:} \quad Z(\bar{x}, \bar{u}),$$

$$\text{subject to:} \quad g_i(\bar{x}, \bar{u}) \leq 0, \qquad i \in 1, .. , l;$$

$$x_{l,j} \leq x_j \leq x_{u,j}, \qquad j \in 1, .. , m; \qquad (6.1)$$

$$u_{l,k} \leq u_k \leq u_{u,k}, \qquad k \in 1, .. , n;$$

in which:
$Z(\bar{x}, \bar{u})$ : objective function;
$g_i(\bar{x}, \bar{u})$ : constraints;
$\bar{x}$       : vector of state variables;
$\bar{u}$       : vector of control variables;
$x_{l,j}, x_{u,j}$ : lower and upper limits on state variables $x_j$;
$u_{l,k}, u_{u,k}$ : lower and upper limits on control variables $u_k$.

Linear Programming assumes a convex piece-wise linear objective function $Z$ and the constraints $g_i$ of Eq. 6.1 to be linear. In linear form, $Z$ and $\bar{g}$ are given by:

$$Z(\bar{x}, \bar{u}) = \bar{p}^T \begin{bmatrix} \bar{x} \\ \bar{u} \end{bmatrix}$$

$$\qquad (6.2)$$

$$\bar{g}(\bar{x}, \bar{u}) = A \begin{bmatrix} \bar{x} \\ \bar{u} \end{bmatrix} - \bar{b}$$

The $(m+n)$-vector $\bar{p}$ contains the penalties corresponding to the state and decision variables and $\bar{p}^T$ denotes the transpose of this vector. $A$ is a real-valued $l \times (m+n)$-matrix and the $l$-vector $\bar{b}$ contains the external inflows and outflows to and from the boundaries of the water system over time at each subsystem.

The basis for the model are the continuity equations for flow and pollution. The structure of this Network Programming problem yields a sparse matrix $A$ in Eq. 6.2. The addition of linear side constraints resulting from linearized process descriptions, transforms this problem into an LP problem. The matrix $A$ however still remains sparse.

## 6.2.2 Forward Estimating

In linear modeling, much effort is generally spent in determining an accurate representation of nonlinear relationships. Nonlinear relationships have to be linearized, preferably in such a way that the linearization process does not violate the original nonlinear behavior.

In the control problem defined, it is most likely that the system state at the end of the control horizon is quite different from the system state at the beginning. Therefore, linearizing at the initial values of the state variables would introduce large inaccuracies with respect to the description of the water-system behavior.

The problem of inaccurate linearization can be dealt with in SLP by making use of the cyclic character of the optimization method. In each cycle, the results of the previous one can be used:

$$\bar{v}^{\tau+1}(t) \approx \bar{v}^{\tau}(t+1), \tag{6.3}$$

in which:
$\bar{v}$     : vector of optimization variables;
$t$     : optimization time step (-), $t \in \{0, 1, .., T\}$;
$\tau$     : simulation time step (-);
$T$     : control horizon (-).

This ensures linearization of relationships at values for variables that are most likely the results of a next optimization. As will be shown in Sec. 7.1.6, this cyclic process can increase the accuracy of the linear model dramatically. The phenomenon is introduced in the present study as *forward estimating*.

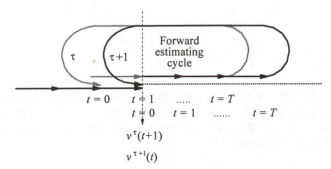

*Fig. 6.2. The principle of forward estimating applied to optimization variable v.*

Figure 6.2 shows the principle of forward estimating. In order to calculate the linearization of the nonlinear constraints, a first order Taylor approximation is determined at the value of the optimization variables of the previous cycle. In equation form this means that the nonlinear function $f$ of the vector $\bar{v}^{\tau+1}(t)$ is linearized at the value of the corresponding vector of the previous cycle, $\bar{v}^{\tau}(t+1)$:

$$f(\bar{v}^{\tau+1}(t)) \approx f(\bar{v}^{\tau}(t+1)) + \bar{\nabla}f(\bar{v}^{\tau}(t+1))^{\mathrm{T}} \cdot (\bar{v}^{\tau+1}(t) - \bar{v}^{\tau}(t+1)), \tag{6.4}$$

in which:
$\bar{v}$        : vector of optimization variables;
$f$        : function of $\bar{v}$;
$\bar{\nabla}$        : nabla vector, $\bar{\nabla}^T = [\partial/\partial v_1, \partial/\partial v_2, .. , \partial/\partial v_n]$.

The principle of forward estimating can be used in the description of water-quantity and water-quality processes. An example of the latter is the transport of pollutants between subsystems, while the direction of flow is not known in advance. If the actual water-system load does not differ from the one predicted for the control horizon, the flow and its direction are known exactly from a previous cycle. In practice, the system load may change as soon as more information becomes available at simulation time $\tau$. For this reason, forward estimating requires small and gradual changes in system load and thus a limited size of optimization time step $\Delta t$.

Forward estimating requires initial estimates at the start of the simulation and optimization loop ($\tau = 0$). Initial estimates can be set fixed or can be estimated by use of the simulation model during the time steps $t \in \{0, 1, .. , T\}$, using an estimate of the Control Strategy.
    Furthermore the forward results at the end of the control horizon $t = T$ are not available during successive forward estimating cycles, therefore the data of $T - 1$ can be used as an estimate for $T$:

$$f(\bar{v}^{\tau+1}(T)) \approx f(\bar{v}^\tau(T)). \tag{6.5}$$

### 6.2.3  Interests Modeled in the Objective Function

A water system can be divided into areas, structures and flow elements, as shown in Fig. 6.3. Each area can consist of a number of subsystems e.g., groundwater, surface water, urban drainage. The objective of water-system control is to satisfy the requirements of all interests present in that water system. The satisfaction of these interests, as described in Sec 3.1, can be expressed in water-system variables such as groundwater levels and pollutant concentrations.
    The requirements of each interest can be incorporated in the objective function $Z$ (Eq. 6.2) by damage functions $D$ of system state variables $x$. Each damage function penalizes the deviation of the state which is required by an interest in a subsystem. This approach is known as Nonpreemptive Goal Programming (Sec. 3.4). The objective function for one time step $\Delta t$ of the control horizon, can be written as:

$$Z(\bar{x}, \bar{u}) = \sum_{i=1}^m W_{ai} \sum_{j=1}^n R_{di,j} D_{i,j}(x) + \sum_{k=1}^r W_{rk} u_k. \tag{6.6}$$

In Eq. 6.6, $D_{i,j}(x)$ represents damage function $j$ for water subsystem $i$. Coefficient $W_{ai}$ represents the weight of an area in the objective function. Coefficient $R_{d\,i,j}$ expresses the relative importance of interest $j$ in water subsystem $i$ within a particular area.

Additional penalties for operations are described by the last part of Eq. 6.6. Through coefficients $W_{r\,k}$ the use of regulating structures can be influenced. This option can be used to create a threshold for the operation of structures and prevent unnecessary operations. These coefficients could be used as well to determine operational sequences of regulating-structure units.

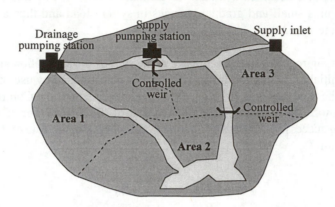

*Fig. 6.3. Example water system with areas, regulating structures and flow elements.*

### 6.2.4  Use of Damage Functions

The value of some interests is hard to determine objectively. Comparison of the output of agricultural produce and the value of aquatic life in a water system, presents a serious rating problem. To enable weighing without having to determine the costs of damage explicitly, a dimensionless objective function has been developed. Determining the damage and the weighing of interests, is a process, which is discussed in Chapter 7.

Damages can be assigned to *key variables*, e.g.: water levels; pollutant concentrations. Graph (a) of Fig. 6.4 shows an example of a damage function of a water-system key variable $x_k$.

One or more damage functions can be used to represent an interest. The damage function incorporates a penalty $p$ or a negative gain to the overall objective of control. The penalty is zero if a subsystem is in its desired state and it increases with deviations from the desired state.

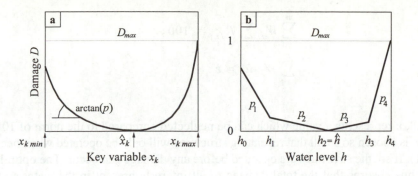

*Fig. 6.4. Nonlinear damage function (a) and its linearized form (b).*

Each damage function incorporated in the objective function (Eq. 6.6) has to be piece-wise linear and convex to permit the use of LP. Graph (b) of Fig. 6.4 shows schematically how a damage function can be linearized to meet the required form. Depending on the required accuracy with which the nonlinear-damage function is given, more or less line segments in the linearized function are required.

In order to keep the objective function dimensionless, penalties in the objective function have the inversed dimension of the key variables they are assigned to. From the damage function the following relationship can be abstracted:

$$p = \frac{dD}{dx_k}$$ (6.7)

in which:
$p$     : penalty (e.g. $m^{-1}$);
$D$     : dimensionless damage (-);
$x_k$    : key variable (e.g. m), $x_k \in [x_{k\,min}, x_{k\,max}]$.

The damage function reflects the harm to, or loss of a particular interest. To enable weighing of interests, represented by damage functions, the damage is considered to have reached its maximum at one of the limits of a pre-defined range of a key variable ($x_{k\,min}$ or $x_{k\,max}$). Comparison of unequal interests is further enabled by fixing the maximum damage of each interest at maximum $D_{max}$, which equals unity.

To enable weighing of unequal interests, a mechanism is used to determine the mutual weights of interests in a water system. Comparison of water systems is possible by introducing a maximum damage for an entire water system during one step of the control horizon, which equals 100. In modeling this can be accomplished by the following equations (see also Eq. 6.6):

$$\sum_{i=1}^{m} W_{ai} \sum_{j=1}^{n} R_{di,j} = 100 \, ,$$

$$\sum_{k=1}^{r} W_{rk} u_k = \varepsilon \, . \tag{6.8}$$

In Eq. 6.8, $\varepsilon$ represents a value which can be neglected compared to the norm of 100. The value of $\varepsilon$ is chosen so small that regulating structures will only be operated when necessary (e.g. 0.01). If so, the structures are operated before any damage can occur. The optimization process thus ensures that the total damage to all interests present in the water system is minimal.

Figure 6.5 presents imaginary examples of piece-wise linear damage functions for several interests. The linearization of a strongly nonlinear damage function can be accomplished by defining several linear line segments, under the restriction that the function is convex.

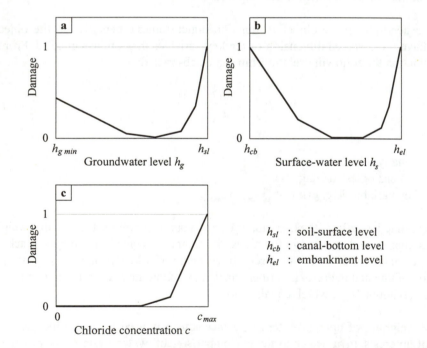

*Fig. 6.5. Example damage functions for groundwater level (a), surface-water level (b) and chloride concentration (c).*

### 6.2.5 Runoff Modeling

The optimization model is based on the simulation model and a further schematization of the real-world water system. Chapter 4 discussed the modeling methods used to simulate the water-system behavior.

Figure 6.6 shows the various flows to and from a groundwater and a surface-water subsystem. The combination of these subsystems can be used to model the five types of real-world subsystems: urban drainage, urban surface water, urban groundwater, rural surface water and groundwater (Fig. 1.3, Page 2).

In the following text, modeling of sewer subsystems is not explicitly mentioned. However, sewer systems can be modeled in the same manner as surface-water subsystems.

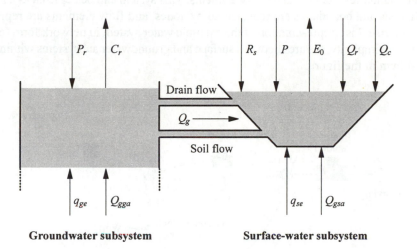

Fig. 6.6. *Groundwater and surface-water flows modeled in optimization.*

The model incorporates the following directly controlled flows:
- flows via regulating structures: pumping stations, weirs, sluices and inlets ($Q_r$).

The model furthermore incorporates the following indirectly controlled flows, resulting from differences in groundwater and surface-water level:
- groundwater flows between the groundwater and surface-water subsystems of different areas ($Q_{gga}$ and $Q_{gsa}$), e.g. regional groundwater flow and shallow seepage;
- surface-water flows between surface-water subsystems of different areas, i.e. flows through canals ($Q_c$).

Finally, the model incorporates the following uncontrolled flows:
- percolation and capillary flows from and into the unsaturated zone ($P_r$ and $C_r$);
- external groundwater flows to and from groundwater and surface water subsystems ($q_{ge}$ and $q_{se}$), e.g. deep seepage or infiltration;

- runoff to surface water from various types of surfaces ($R_s$), e.g.: runoff from semi-impermeable surfaces, runoff from uncontrolled sewer systems;
- precipitation into and evaporation from surface waters ($P$ and $E_0$).

The flows which are controlled directly or indirectly, are incorporated in the optimization problem as variables, the uncontrolled internal and external flows are incorporated as boundary conditions. All boundary conditions are determined by the prediction module.

### 6.2.6  Network Model

As described in Sec. 4.1, a water system can be represented schematically by a system of areas and boundaries, connected by flow elements. This system can be depicted as a network, in which areas and boundaries are represented by nodes, and flow elements are represented by arcs. Figure 6.7 is a representation of the example water system in network form (compare Fig. 6.3). For simplicity, the arcs between surface and groundwater subsystems within an area are not shown in the figure.

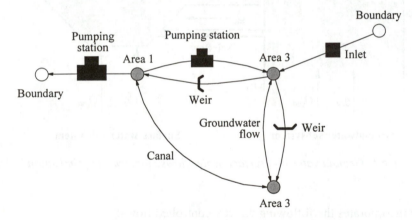

*Fig. 6.7. Example water system in network form.*

In this network, water can be transported from one area to another via the intersecting arcs. Areas of the water system are modeled in the network by nodes.

The network in Fig. 6.7 describes flows in a static situation. Introducing time-varying flows to and from nodes, the network expands in time, since for each time step ($t = 1, 2, .. , T$) the network of Fig. 6.7 is replicated. Figure 6.8 shows such a network. Note that the situation at $t = 1$ is known from simulation and is thus not included in the network.

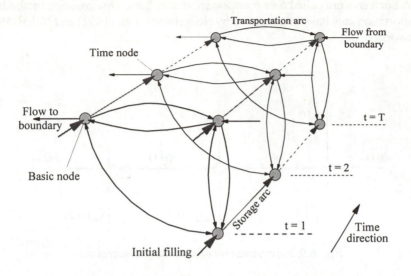

*Fig. 6.8. Water-system network expanded in time.*

In Fig. 6.8 the nodes at $t = 1$ are called *basic nodes* since they represent the original areas of the water system and form the basis of the node structure in successive time steps. The nodes of successive time steps are called *time nodes*.

At the basic nodes, the initial states of the areas of the water system are modeled as inflows, e.g. the initial filling of the surface-water subsystems. The nodes are connected by *transportation arcs* within a time step and by *storage arcs* between time steps. The transportation arcs represent flows through flow elements during a time step, the storage arcs represent the storage in subsystems.

Figure 6.9 presents the basic structure of the mass balances for the time nodes. In the optimization model these balances are described by water quantity and water quality *continuity constraints*, in general format:

$$v(t) = v(t-1) + v_{in}(t) - v_{out}(t), \tag{6.9}$$

in which:
| | | |
|---|---|---|
| $v$ | : | water-system variable; |
| $v_{in}$ | : | inflow into a node; |
| $v_{out}$ | : | outflow from a node; |
| $t$ | : | optimization time step. |

It should be mentioned that the transportation arcs described above represent flows *at* a certain time (*time-independent transportation arcs* in Fig. 6.10).

In principle, it is possible to model arcs that include a time component in flow, where the size of a time step determines the time required for a wave to flow from one subsystem to another. Such arcs are called *time transportation arcs* here. Optimization methods, using time transportation arcs have been applied by Neugebauer et al. (1991) and Nelen (1992) for sewer networks.

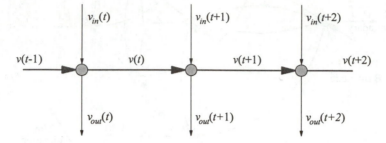

*Fig. 6.9. Representation of continuity constraints.*

Time transportation arcs can present an inherent problem if the penalties assigned to flows through these arcs are lower than the penalties assigned to flows through storage arcs. The problem emerges if a particular route through the network from $t = 1$ to $t = T$, can be chosen, which consists of time transportation arcs only. During system loads, which result in undesirable storage of excess water or pollutants in subsystems, it would be advantageous in optimization to steer all excess storage flow via time transportation arcs instead of via storage arcs. Time transportation arcs thus keep the undesirable substances in transit, in fact removing these substances from the modeled water system.

An example of such a phenomenon is a two-area water system, with a connecting pumping station that discharges in one direction and an inlet that discharges in the opposite direction, shown schematically in diagram (b) in Fig. 6.10. The optimal flows through the network will always choose the route with the lowest penalty. During periods of excessive water or pollution inflow, the optimization result shows the target storages in the subsystems, combined with a constant flow through the pumping station and the inlet.

Decisive for the phenomenon mentioned, is whether the optimization model includes penalties for transportation and whether these are lower than the penalties for storage. In general, the penalties for transportation via structures are much lower than the penalties for storage, since the structures are specifically intended to be used if undesirable substances enter the water system. In practice, the phenomenon mentioned will show up when water systems that contain both drainage and supply structures are modeled. This situation exists in the majority of water systems in the Netherlands.

In conclusion, time transportation arcs can only be used in optimization models of water systems if the flow is uni-directional. Some sewer systems can be schematized in this way or actually even have the required layout. However, looped networks cannot be modeled using time transportation arcs. The present study focuses on the general water system and therefore time-independent transportation arcs have to be used.

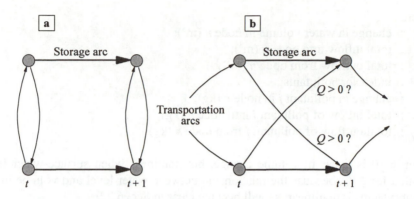

*Fig. 6.10. Time-independent transportation arcs (a)*
*and time-dependent transportation arcs (b).*

### 6.2.7 Two Modeling Approaches

In mathematical programming, the way in which a model is built up and the way in which penalties are assigned to state and control variables, are decisive for increasing the speed in finding the optimal solution and for the flexibility available in modeling.

In the Constraint Manager, several mathematical modeling approaches for describing the physical water system can be used. In-depth discussion and performance considerations with respect to these approaches can be found in Hoogendoorn (1996).

Two approaches are discussed here: a generalized network approach with side constraints, called 'generalized network' and a general-LP approach, called 'general LP'. The following discussion focuses on building the network and how damage functions can be included in the optimization model.

#### *Generalized Network*

This approach formulates the optimization problem as a generalized network problem, with additional side constraints for modeling physical water-system processes (Sec. 3.4.2). To maintain a pure network structure whenever possible, all variables within a network must have the same dimensions.

The pure network-type equations follow from the water-quantity and water-quality continuity equations for each node:

$$\Delta V_n = \sum Q_{n\,in} - \sum Q_{n\,out}, \qquad (6.10)$$

$$\Delta K_{nj} = \sum K_{n\,in\,j} - \sum K_{n\,out\,j}, \qquad (6.11)$$

in which:

$\Delta V_n$ : change in water volume in node $n$ (m³);
$\sum Q_{n\,in}$ : total inflow into node $n$ (m³);
$\sum Q_{n\,out}$ : total outflow from node $n$ (m³);
$j$ : index for pollutants;
$\Delta K_{nj}$ : change in pollutant $j$ in node $n$ (kg);
$\sum K_{n\,in\,j}$ : total inflow of pollutant $j$ into node $n$ (kg);
$\sum K_{n\,out\,j}$ : total outflow of pollutant $j$ from node $n$ (kg).

Equation 6.10 is linear in volume change, but nonlinear when surface-water levels are substituted for $\Delta V_n$, because the relationship between water level and volume in surface-water subsystems is nonlinear, as will become clear in Sec. 6.2.9.

In the generalized network a time-expanded network as shown in Fig. 6.8, should be constructed for the water quantity model and also for the model of each water-quality variable. Pollutants are inherently transported with water that flows through network arcs. This can be modeled by means of 'network linking' of water-quantity and water-quality networks, using the equations:

$$K_{i,j} = c_{i,j} Q_i,$$
$$c_{i,j} = c_{a\,i,j}, \qquad (6.12)$$

in which:

$K_{i,j}$ : flow of pollutant $j$ through arc $i$ (kg);
$c_{i,j}$ : concentration of pollutant $j$ in arc $i$ (kg/m³);
$Q_i$ : water quantity flowing through arc $i$ (m³);
$c_{a\,i,j}$ : concentration of pollutant $j$ in the abstraction node of arc $i$ (kg/m³);
$i$ : arc number;
$j$ : index for pollutants.

This linking equation is quadratic and therefore a linearized approximation should be added to the optimization model as a side constraint. A specific problem of Eq. 6.12 is that it depends on variables of which the behavior has already been described by linearized equations and that it requires additional linearization. Such linearization can be achieved, but, as proven by Vredenberg (1996), the outcome of the optimization process very much depends on the accuracy of the predicted system load.

For all water-quantity and water quality-variables which have to be controlled, a piecewise-linear convex damage function has to be defined in terms of volume or mass, e.g. a water level in the surface-water subsystem. One damage function can be imposed on each storage arc (Fig. 6.11) in the network model.

Suppose the damage function given in Fig. 6.11 is imposed on a storage arc of the network (a). The limits on the flow $V$ through this arc are given by $[V_0, V_4]$. On the basis of the shape of the damage function, the target volume $\hat{V}$ corresponds to volume $V_2$ and a deviation from this target is penalized with penalties $p_i$.

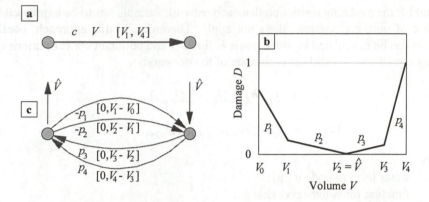

*Fig. 6.11. Network modeling of a piecewise-linear damage function: storage arc (a), linear damage function (b), storage-arc representation in subarcs (c).*

The penalties $p_i$ depend on the value of $V$. Two methods can be distinguished which incorporate the penalties of the nonlinear damage function in the model. The first method is applied by Nelen (1992) and consists of estimating the most likely value of $V$, and determining the penalty for each time step of the control horizon. This method may result in rather inaccurate optimization results when the optimization step or the control horizon is large. Furthermore, the method does not guarantee the optimization process to produce unique solutions in successive optimization steps.

The second method consists of splitting up each storage arc into separate subarcs, each of which has its own penalty and associated capacity. When this method is applied, each linear segment of the damage function has its own storage arc in the network model (c in Fig. 6.11). Again, several methods for implementing this technique are possible. A generally applicable method is the method in which the damage function is shifted to the origin. This method prevents inaccurate results in case $V$ is small in comparison to $V_4 - V_0$ (Roos, 1987).

In the application of the method, the target volume $\hat{V}$ is abstracted from each time node and added to the following time node. This implies an offset of the horizontal axis by $\hat{V}$. Generally, network-flow problems are formulated in such a way that all flows through the network are positive. To guarantee positive flows, the subarcs corresponding to the ascending line segments in the damage function have the same direction as the original storage arc, while the descending line segments have the opposite direction.

It should be noted that the decreasing nature of the first linear line segment implies that the penalty for this section is negative (e.g. $p_1 < 0$). A damage function that has a horizontal line segment presents a subarc with a penalty of zero.

### General LP

In general LP the modeling restriction that each network variable has to be expressed in the same type of unit, e.g. volume, does not apply. Therefore, in this approach, continuity constraints can be formulated in water levels $h$, flows $Q$ and pollutant concentrations $c$. The following equations are valid for each node of the network:

$$\Delta h_n = g_h(h_n, \Sigma Q_{n\ in}, \Sigma Q_{n\ out}), \tag{6.13}$$

$$\Delta c_{n,j} = g_c(h_n, c_{n,j}, c_{n\ in\ i,j}, Q_{n\ in\ i}), \tag{6.14}$$

in which:

| | |
|---|---|
| $h_n$ | : water level in node $n$ (m); |
| $g_h$ | : function for water-level change; |
| $\Sigma Q_{n\ in}$ | : total inflow into node $n$ (m$^3$/s); |
| $\Sigma Q_{n\ out}$ | : total outflow from node $n$ (m$^3$/s); |
| $i$ | : index for in- and outflowing arcs; |
| $c_{n,j}$ | : concentration of pollutant $j$ in node $n$ (mg/l); |
| $g_c$ | : function for change in pollutant-concentration; |
| $c_{n\ in\ i,j}$ | : inflow concentration of pollutant $j$ into node $n$ via arc $i$ (mg/l); |
| $Q_{n\ in\ i}$ | : total inflow into node $n$ via arc $i$ (m$^3$/s); |
| $j$ | : index for pollutants. |

In general LP, water-quantity and water-quality networks are linked by means of the following equation:

$$c_{in\ i,j} = c_{a\ i,j}, \tag{6.15}$$

in which:

| | |
|---|---|
| $c_{in\ i,j}$ | : concentration of pollutant $j$ transported by arc $i$ to its delivery node (mg/l); |
| $c_{a\ i,j}$ | : concentration of pollutant $j$ in the abstraction node of arc $i$ (mg/l); |
| $j$ | : index for pollutants. |

This linking equation is linear and can be substituted in the continuity equation for pollution (Eq. 6.14) directly.

For each state variable $x$ a piecewise-linear convex damage function can be defined. Suppose the damage function of Fig. 6.12 has a target value of $x = \hat{x}$. This damage function can be described in a very flexible way, by defining line segments one to four and imposing the solution of the optimization process to produce a water-level-damage combination which is above or on each of the lines.

To apply this technique, an additional optimization variable $\eta$ is required in the objective function (Eq. 6.6) for each damage function $D(x)$. The purpose is to force the solution to be on the damage curve.

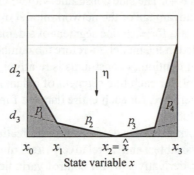

Fig. 6.12. General LP modeling of a piecewise-linear damage function.

The line segments of the damage function can be described by functions $l_j$ of variable $x$, in combination with the restriction that the solution has to be above or on a line, the following inequality is obtained:

$$l_j(x) = p_j x + d_j \leq \eta,$$                    (6.16)

in which:

$l_j$      : linear function describing damage function line segment $j$;
$x$      : water-system state variable;
$p_j$      : penalty for damage-function line segment $j$;
$d_j$      : damage at intersection of $l_j$ with the vertical axis ($x = 0$);
$j$      : index for line segments;
$\eta$      : damage variable (-), generally $\eta \in [0, 1]$.

The optimization variable $\eta$ should be scaled in such a way that it ensures a solution of the damage function and does not interfere with other variables. For scaling purposes the weighing coefficient of this variable in the objective function can be set to the arbitrary value of unity. Interference with other variables can be excluded by using the $\eta$-variables for determining the damage to state variables only.

When a range is defined in the damage function for which state variable $x$ has no damage, a horizontal line develops. The constraint in the optimization model, which describes such a line, is unnecessary and can thus be removed.

## 6.2.8 Choice of Approach

Hoogendoorn (1996) made an inventory of optimization solvers and their applicability in water-system control. He found that formulation of the problem as a generalized network or as general LP, has a rather large effect on the solution speed of various solvers.

Building models of water systems as a generalized network produces a relatively large number of side constraints. The side constraints result from descriptions of water-transport relationships (Sections 6.3 and 6.4). The side constraints violate the network structure of the problem, and thus reduce the advantage of the network model over other types of models.

In the generalized network, for each line segment of a damage function, an additional variable $V$ has to be incorporated in the model, therefore the number of variables is relatively large. However, the number of continuity constraints is relatively low.

In the general LP model, for each line segment of a damage function, an equation $l$ is included in the model. Furthermore, for each entire damage function, only one additional optimization variable $\eta$ is included.

The generalized network produces a compact optimization problem, comprising a relatively high number of variables and a relatively low number of constraints. The generalized network requires derivatives of all control variables (e.g. water level) to be incorporated in the model by means of storages. This requires extra linearizations each forward estimating cycle. In general LP this is not required.

Both approaches require scaling of variables and equations in such a way that the problem can be solved accurately. Scaling of variables is very complex for a generalized network and, in fact, depends on the properties of the modeled water-system. The phenomenon that the storage in each node has to be transported in time through storage arcs, can result in variables of which the optimal values are not of the same order of magnitude. An example is the difference in scale when modeling the storage in a groundwater subsystem (e.g. 100,000 ha) together with the storage in a much smaller surface water-sub system (e.g. 10 ha). In general LP, water quantity can be modeled by means of water levels in subsystems, which are always of the same order of magnitude and require no additional scaling.

Combined modeling of water quantity and quality considerably enlarges the generalized network model with complex side constraints that describe the linearized relationship between flow and pollutants (Eq. 6.12). These equations are based on linearized relationships and have to be linearized again. As a result, the solution of the optimization process depends strongly on the accuracy of the system-load prediction. In general LP, the inclusion of water quality is less complex and the water-quality linking equation (Eq. 6.15) can even be substituted in the continuity equations for water quality (Eq. 6.14), effectively removing them from the optimization problem.

According to Hoogendoorn (1996), the network model with side constraints can be solved faster than the general LP model, using an LP solver. Solution speed strongly depends on the solver algorithm, though. His general conclusion is that solvers of which the algorithm is based on the Interior Point method (Terlaky, 1996) perform considerably better than solvers which are based on the Simplex method (Hillier & Lieberman, 1990), for both modeling approaches. Speed improvements by a factor four are reported for Interior Point methods in comparison with Simplex methods when applied to water-control problems. For this reason, an Interior Point method seems the obvious choice.

If an Interior Point method is used, the solution speed of a generalized network problem is slightly better than that of general LP (approximately 15%). Combined with the described difficulties in formulating an efficient generalized network, general LP has been preferred in the present study. Table 6.1 shows a comparative overview of the above discussion.

Table 6.1. Comparison of two approaches for building the optimization problem.

| Approach | Generalized network | General LP |
|---|---|---|
| Type of optimization variables | Volume, mass | As occur in process descriptions |
| Units | $m^3$ and kg | mm, mm/h, m, mg/l, etc. |
| Characteristics optimization matrix $A$ | More columns (variables) than rows (constraints) | Generally more rows than columns |
| Types of constraints | Mainly equalities | Mainly inequalities |
| Water quantity and quality linking | Requires additional linearizations, can produce unstable results | Requires no additional constraints or linearizations |
| Scaling of variables | Complex, depends on network | Can be generally determined |
| Scaling of equations | Complex, in combined modeling of water quantity and water quality | Less complex, similar for water-quantity and water-quality modeling |
| Extendability of the model | Complex | Relatively easy |
| Problem solving speed using an interior point method | Fast | Slightly less fast than generalized network |

## 6.2.9  Frequently Used Relationships

In the following sections, some relationships are used frequently. Therefore they are summarized and generally described here. It should be mentioned once more, that modeling of sewer systems is not described explicitly, but is similar to surface-water modeling.

Figure 6.13 shows several properties of a surface-water subsystem. Four functions are used to describe the relationship between surface-water level, surface-water area and surface-water volume in the optimization model.

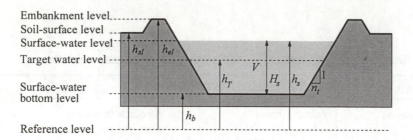

*Fig. 6.13. Generally used terms in a surface-water subsystem.*

The 'volume function' $f_V(h)$, which depends on the water level $h$ in a subsystem:

$$f_V(h) = [A_T + n_t L (h - 2 h_T + h_b)] (h - h_b).  \tag{6.17}$$

'Area functions' $f_A(V)$ and $f_A(h)$, which depend on the volume $V$ of a subsystem and the associated water level $h$ respectively:

$$f_A(V) = [A_T^2 + 4 n_t L (V - V_T)]^{\frac{1}{2}},  \tag{6.18}$$

$$f_A(h) = A_T + n_t L (2 h - 2 h_T + h_b).  \tag{6.19}$$

The 'water-level function' $f_h(V)$, which depends on the volume $V$ of a subsystem:

$$f_h(V) = \begin{cases} h_T + \dfrac{f_A(V) - A_T}{2 n_t L} & \text{for } n_t > 0, \\[2ex] \dfrac{V}{A_T} & \text{for } n_t = 0, \end{cases}  \tag{6.20}$$

in which:

$f_V(h)$ : volume function depending on water level (m³);
$f_A(V)$ : area function, depending on volume (m²);
$f_A(h)$ : area function, depending on water level (m²);
$f_h(V)$ : water-level function, depending on volume (m+ref);
$h_T$ : target water level (m+ref);

$A_T$ : area at target level (m$^2$);
$V_T$ : volume at target level (m$^3$);
$n_t$ : total side slope of the subsystem (-);
$L$ : total length in the surface-water subsystem (m);
$h_b$ : surface-water bottom level (m+ref).

The derivatives of the water-level function and the volume function are also used frequently in the linearizations described in the next sections. To reduce the complexity of the equations, on the basis of Eq. 6.17, 6.18, 6.19 and 6.20 the following relationships are deduced:

$$\frac{df_h}{dV}(V) = \frac{1}{f_A(V)},$$

(6.21)

$$\frac{df_V}{dh}(h) = f_A(h).$$

(6.22)

### 6.2.10 Conventions Used

In the following sections, several of the relationships presented in Chapter 4 are translated into constraints of the optimization model. Only the relationships required in optimization are described further.

A general, distinction has been made between variables, parameters and other data:

- 'optimization variables' are explicitly included in the optimization model as control and state variables;
- 'forward-estimated parameters' are model coefficients of which the value depends on the value of optimization variables in a previous forward estimating cycle at the same moment in time, in case they represent values of linearized variables, forward estimated parameters are indicated by '*';
- 'model parameters' are parameters as used in simulation;
- 'time-series data' are data which are produced by the prediction model on the basis of a predicted water-system load.

As mentioned before, scaling of the optimization problem is very important to maintain a high solving speed. To scale optimization variables, different units have been used for the constraints in the model, e.g.: water level (m+ref), pollutant concentration (mg/l), flow (m$^3$/s). For this reason, conversion constants are included in the equations, which are described in the following sections.

The general optimization problem built, including the objective function and its constraints, is summarized at the end of this chapter.

## 6.3  Subsystem Modeling

### 6.3.1  Groundwater

A groundwater system consists of a saturated and an unsaturated zone. In optimization only the saturated zone is incorporated. The prediction model provides an estimate of the interactions between the saturated and the unsaturated zones. In the approach chosen, it is assumed that the mutual influence between these zones as a result of control actions only, is small during the control horizon and can thus be neglected.

The resulting difference between percolation $P_r$ and capillary rise $C_r$ is taken as the boundary condition for the continuity constraint of the saturated zone. Figure 6.14 shows the groundwater subsystem as modeled in optimization. Note that the groundwater subsystem does not have a bottom level.

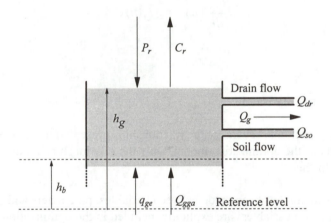

*Fig. 6.14. Groundwater subsystem as modeled in optimization.*

The modeled groundwater subsystem exchanges water with its environment and interacts with:

- the surface-water subsystem of the same modeled area ($Q_g$);
- uncontrolled external groundwater ($q_{ge}$);
- the surface-water or groundwater subsystem of another area ($Q_{gga}$).

In this section, the interaction between the groundwater and surface-water subsystems of one area as well as the external in- and outflows will be discussed. Section 6.4.6 describes the flows between groundwater and/or surface-water subsystems in different areas ($Q_{gga}$).

The groundwater flow between the groundwater and the surface-water subsystems within an area is called 'groundwater discharge' $Q_g$ in the following. Two types of groundwater discharge are distinguished: discharge through the soil and discharge through drains. The latter can be used to model physical drains in the soil or to model surface runoff in rural areas.

For each groundwater subsystem two types of constraints are distinguished: continuity constraints and side constraints.

### Continuity Constraint

The continuity constraint for the groundwater subsystem as shown in Fig. 6.14 yields:

$$h_g(t) \; = \; h_g(t-1) + \frac{1}{\mu^*} \left[ -Q_g(t) + \sum Q_{gga}(t) + \frac{P_r^*(t) - C_r^*(t) + q_{ge}(t)}{3600 \cdot 1000} \right] \Delta t , \quad (6.23)$$

in which the following are:
optimization variables:
$h_g$      : groundwater level (m+ref), $h_g \in (-\infty, \infty)$;
$Q_g$      : groundwater discharge (m³/s), $Q_g \in (-\infty, \infty)$;
$\sum Q_{gga}$      : inflow via groundwater flow elements from other areas (m³/s);
model parameters:
$t$      : optimization time step (-);
$\Delta t$      : size of optimization time step (s);
forward-estimated parameters:
$P_r^*$      : percolation in the previous cycle (mm/h), $P_r \in [0, \infty)$;
$C_r^*$      : capillary rise in the previous cycle (mm/h), $P_r \in [0, \infty)$;
$\mu^*$      : effective storage coefficient (-), $\mu^* \in (0, 1]$;
time-series data:
$q_{ge}$      : uncontrolled external flow to the groundwater subsystem (mm/h), $Q_{ge} \in (-\infty, \infty)$.

If the water level in the groundwater subsystem rises above the soil-surface level, the effective storage coefficient of the soil is not valid anymore. The continuity constraint Eq. 6.23 presents the change in water level and therefore it can also be used to describe water-level changes above the soil surface. In that case, the effective storage coefficient $\mu$ is fixed to unity and the external boundary flows of percolation and capillary rise (Fig. 6.14) are replaced by precipitation and evaporation. This approach enables the formation of puddles on the surface, whereas impossible rises in groundwater level are prohibited.

### Side Constraint

The groundwater and surface-water subsystems interact via the groundwater discharge. If the groundwater level is below the level of the drains, only slow discharge through the soil takes place. If the groundwater level rises above the level of the drains, additional rapid discharge occurs through those drains.

A distinction is made between outflow from the soil and inflow into the soil. This is required to incorporate the difference in resistance coefficients for soil out- and inflow. The binary forward-estimated parameters $\kappa_d$ and $\kappa_s$ on the basis of the groundwater and surface-water levels, determine the medium of flow: through the soil only or through soil and drains:

$$\kappa_d = \begin{cases} 1, & \text{if } h_g^* \geq h_{dr}, \\ 0, & \text{if } h_g^* < h_{dr}, \end{cases} \tag{6.24}$$

$$\kappa_s = \begin{cases} 1, & \text{if } h_s^* \geq h_{dr}, \\ 0, & \text{if } h_s^* < h_{dr}. \end{cases} \tag{6.25}$$

The total groundwater discharge $Q_g$ is the summation of the slow discharge through the soil $Q_{so}$ and rapid discharge through drains $Q_{dr}$. The resulting side constraint for both discharges together can be described on the basis of Eq. 4.1 (Page 69) by:

$$Q_{so} = \frac{A_g (h_g - h_s)}{S_{rg}},$$

$$Q_{dr} = \frac{A_g [\kappa_d h_g - \kappa_s h_s + (\kappa_s - \kappa_d) h_{dr}]}{S_{rd}},$$

$$Q_g = Q_{so} + Q_{dr} \quad \Leftrightarrow \tag{6.26}$$

$$Q_g = A_g \left[ \left( \frac{1}{S_{rg}} + \frac{\kappa_d}{S_{rd}} \right) h_g - \left( \frac{1}{S_{rg}} + \frac{\kappa_s}{S_{rd}} \right) h_s + \frac{(\kappa_s - \kappa_d)}{S_{rd}} h_{dr} \right],$$

in which the following are:
optimization variable:
$Q_g$     : groundwater discharge (m³/s);
forward-estimated parameters:
$\kappa_d$     : binary parameter for flow medium (-);
$\kappa_s$     : binary parameter for flow medium (-);
model parameters:
$A_g$     : groundwater subsystem area (m²);
$S_{rg}$     : soil resistance (s);
$S_{rd}$     : drain resistance (s);
$h_{dr}$     : drain level (m+ref).

The water levels $h_g$ and $h_s$ are available as forward-estimated parameters each cycle. Inaccurate determination of $h_g$ and $h_s$ may lead to wrong estimates of the direction of flow through soil and drains. This leads to inaccuracies in modeling results, especially for water quality. Without special measures, water can flow with the wrong resistance and quality out of or into the soil. To handle this problem, the 'certainty parameter' $\alpha$ for soil outflow is introduced, which ranges from zero to unity.

If the groundwater level is clearly above the surface-water level and there is only little chance that the flow may revert during a time step of the control horizon, the certainty parameter becomes unity. In the opposite situation, if the groundwater level is clearly below the surface-water level and there is only little chance that the flow may revert, the certainty parameter becomes zero. To approximate this, the following function is used:

$$\alpha_g = \frac{1}{2} + \frac{\arctan\left[a_g\left(h_g^* - h_s^*\right)\right]}{\pi}, \tag{6.27}$$

in which the following are:
forward-estimated parameter:
$\alpha_g$     : certainty parameter for direction of groundwater flow (-), $\alpha_g \in (0, 1)$;
model parameter:
$a_g$     : accuracy parameter ($m^{-1}$), $a_g \in (0, \infty)$.

When the accuracy parameter $a_g$ is increased to infinity, the above equation acts as a step function. If the accuracy parameter is set well, the equation yields a value of almost unity if the groundwater level is well above the surface-water level. A very small value results if the groundwater level is well below the surface-water level.

The value of $S_{rg}$ can finally be defined as the weighted sum of the soil inflow- and outflow resistances:

$$S_{rg} = \alpha_g S_{rg\ out} + (1 - \alpha_g) S_{rg\ in}, \tag{6.28}$$

in which the following are model parameters:
$S_{rg\ out}$     : soil outflow resistance (s);
$S_{rg\ in}$     : soil inflow resistance (s).

The entire groundwater discharge constraint Eq. 6.26 proves to be linear and thus can be incorporated into the optimization problem, without further adaptations.

## 6.3.2 Surface Water

### Water Quantity

Figure 6.15 shows the way in which the surface water subsystem is modeled in optimization. Contrary to the groundwater subsystem, the surface water subsystem does have a bottom level. When the surface water has fallen to the bottom level, it is empty.

Empty subsystems represent a separate type of problem in modeling. In optimization, water levels $h_s$ should not fall below the bottom of the subsystem. This is especially important in modeling surface-water quality.

As will become clear when describing the modeling of regulating structures (Sec. 6.4), a situation may occur where more water is withdrawn from the surface-water subsystem than available. In those cases, to ensure that the water depth $H_s \geq 0$, additional measures have to be taken. Therefore, the surface-water inflow variable $Q_{us}$ is included in the model, which abstracts water from a virtual reservoir. This variable is added to the objective function Eq. 6.6, with a large penalty, to restrict its use to those cases when no other options are available to prevent the surface-water level from falling below the subsystem bottom ($H_s < 0$).

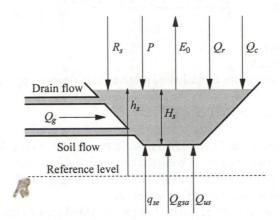

*Fig. 6.15. Surface-water subsystem as modeled in optimization.*

Several physical flows to and from the surface-water subsystem can be distinguished:
- discharge from the groundwater subsystem of the same area $Q_g$;
- discharges from other areas via flow elements ($Q_f$) such as regulating structures $Q_r$, canals $Q_c$ and groundwater flow $Q_{gsa}$;
- external inflows from outside the modeled water system $q_{se}$;
- runoff from various types of surfaces $R_s$;
- precipitation $P$ and surface-water evaporation $E_0$.

For the surface-water subsystem only a continuity constraint is required:

$$\frac{f_V(h_s(t)) - f_V(h_s(t-1))}{\Delta t} = Q_g(t) + \sum Q_f(t) + Q_{us}(t)$$
$$+ \frac{A_s \, q_{se}(t)}{3600 \cdot 1000} - \frac{A_s \, [P(t) - E_0(t)]}{1000 \, \Delta t}, \tag{6.29}$$

in which the following are:
optimization variables:
$h_s$ : surface-water level (m+ref), $h_s \in [0, \infty)$;
$Q_g$ : groundwater discharge (m³/s), $Q_g \in (-\infty, \infty)$;
$\sum Q_f$ : inflow via modeled flow elements (m³/s);
$Q_{us}$ : flow to prevent the surface-water level from falling below the subsystem bottom (m³/s), $Q_{us} \in [0, \infty)$;
model parameters:
$A_s$ : area of surface water (m²);
$t$ : optimization time step (-);
$\Delta t$ : size of optimization time step (s);
time-series data:
$P$ : precipitation (mm);
$E_0$ : surface-water evaporation (mm);
$q_{se}$ : external inflow into the surface-water subsystem (mm/h).

The left-hand side of this equation represents the volume change in the surface-water subsystem per second: $\Delta V_s$. This function is nonlinear and has to be linearized by using a first-order Taylor approximation (Eq. 6.4):

$$
\begin{aligned}
\Delta V_s(h_s(t), h_s(t-1)) \ &= \ \frac{f_V(h_s(t)) - f_V(h_s(t-1))}{\Delta t} \\[2mm]
&= \ \frac{\{[A_T + n_t L(h_s(t) - 2h_T + h_{cb})](h_s(t) - h_{cb})\}}{\Delta t} \\[2mm]
&\quad - \ \frac{\{[A_T + n_t L(h_s(t-1) - 2h_T + h_{cb})](h_s(t-1) - h_{cb})\}}{\Delta t} \\[2mm]
&\approx \ k_{hs1}\,h_s(t) + k_{hs2}\,h_s(t-1) + k_{hs3},
\end{aligned}
$$

(6.30)

in which the following linearization constants represent:

$$
\begin{aligned}
k_{hs1} \ &= \ \frac{\partial \Delta V}{\partial h_s(t)}(h_s^*(t), h_s^*(t-1)) = \frac{\partial f_V}{\partial h}(h_s^*(t)) = f_A(h_s^*(t)), \\[2mm]
k_{hs2} \ &= \ \frac{\partial \Delta V}{\partial h_s(t-1)}(h_s^*(t), h_s^*(t-1)) = -\frac{\partial f_V}{\partial h}(h_s^*(t-1)) = -f_A(h_s^*(t-1)), \\[2mm]
k_{hs3} \ &= \ \Delta V_s(h_s^*(t), h_s^*(t-1)) - k_{hs1}h_s^*(t) - k_{hs1}h_s^*(t-1),
\end{aligned}
$$

(6.31)

in which the following are:
function:
$\Delta V_s$ : surface-water volume-change function (m³/s);
forward-estimated parameter:
$h_s^*$ : surface-water level in the previous cycle (m+ref).

## *Water Quality*

Each water-quality variable included, yields a separate continuity equation in the optimization problem. Pollutants are considered to be transported between surface-water subsystems and between these systems and their surroundings. Pollutants can enter the surface water by inflow, e.g.: pollutants from combined sewer overflows; pollutants from agricultural drainage water.

It is essential to consider that the concentration of water transported by flow elements, only depends on the concentration in the area from which water is abstracted by those flow elements (Eq. 6.15). If a flow reverts, e.g. in a canal, the linked concentration should always switch to the current subsystem from which the water is abstracted. On the basis of Eq. 4.5 (Page 74), the following continuity constraint for water quality can be determined for each surface-water subsystem:

$$\frac{c(t)}{1000\,\Delta t} = [1 - D_{rj}\,\Delta t]\,\frac{c(t-1)}{1000\,\Delta t}$$

$$+ \frac{\sum \dfrac{c_{in}(t)}{1000}\,Q_{in}(t) - \dfrac{c(t)}{1000}\sum [\,Q_{in}(t) + Q_u(t)\,] + \sum K_{uj}(t)}{f_V(h_s(t))} \qquad (6.32)$$

This equation can be rewritten for each area *a*, including the specific concentrations of each area from which water is abstracted by flow elements *j*:

$$\frac{\{c_{aj}(t) - [1 - D_r\,\Delta t]\,c_{aj}(t-1)\}\,f_V(h_s(t))}{1000\,\Delta t} - \sum_{f_{in}} \frac{c_{fj}(t)\,Q_f(t)}{1000}$$

$$+ \sum_{f_{in}} \frac{c_{aj}(t)\,Q_f(t)}{1000} + \frac{c_{aj}(t)\,Q_u(t)}{1000} = \sum K_{uj}(t), \qquad (6.33)$$

in which the following are:
optimization variables:
$c_{aj}$   : concentration of pollutant *j* in a surface-water subsystem (mg/l);
$h_s$   : surface-water level in the subsystem (m+ref);
$Q_f$   : inflow into the surface-water subsystem via flow element *f* (m³/s);
$c_{fj}$   : concentration of pollutant *j* supplied by flow element *f* (mg/l);
$Q_u$   : uncontrolled inflow into the surface-water subsystem (m³/s);
model parameters:
$D_{rj}$   : decay rate of pollutant *j* (s⁻¹);
$\sum K_{uj}$   : uncontrolled inflow of pollutant *j* into surface-water subsystem (kg/s).
$\Delta t$   : optimization time step (s).

This equation is highly nonlinear and should be approximated by a linear function. For linearization purposes, the following functions are defined:

$$K_a(c, h) = \frac{c\, f_V(h)}{1000\, \Delta t},$$

$$K_f(c, Q) = \frac{c\, Q}{1000},$$

(6.34)

in which:

$K_a$ : pollution flux for area $a$ (kg/s);
$K_f$ : pollution flux via flow element $f$ (kg/s).

Substituting these functions, the water-quality continuity constraint (Eq. 6.33) can be rewritten as:

$$K_a(c_{aj}(t), h_s(t)) - (1 - D_{rj}\,\Delta t)\, K_a(c_{aj}(t-1), h_s(t)) - \sum_{f_{in}} K_f(c_{fj}(t), Q_f(t))$$

$$+ \sum_{f_{in}} K_f(c_{aj}(t), Q_f(t)) + K_f(c_{aj}(t), Q_u(t)) = \sum K_{uj}(t).$$

(6.35)

The left-hand side of this equation defines a new function of optimization variables $\bar{v}$ called $K(\bar{v})$:

$$
\begin{aligned}
K(\bar{v}) &= K(c_{aj}(t), h_s(t), c_{aj}(t-1), c_{fj}(t), Q_f(t)) \\
&= K_a(c_{aj}(t), h_s(t)) - (1 - D_{rj}\,\Delta t)\, K_a(c_{aj}(t-1), h_s(t)) \\
&\quad - \sum_{f_{in}} K_f(c_{fj}(t), Q_f(t)) + \sum_{f_{in}} K_f(c_{aj}(t), Q_f(t)) + K_f(c_{aj}(t), Q_u(t)).
\end{aligned}
$$

(6.36)

The first-order Taylor approximation of the function $K(\bar{v})$ forms a continuity constraint in the optimization problem according to:

$$
\begin{aligned}
K(\bar{v}) &= K(c_{aj}(t), h_s(t), c_{aj}(t-1), c_{fj}(t), Q_f(t)) \\
&\approx k_{q1}\, c_{aj}(t) + k_{q2}\, h_s(t) + k_{q3}\, c_{aj}(t-1) + \sum_{f_{in}} [k_{qf1}\, c_{fj}(t) + k_{qf2}\, Q_f(t)] + k_{q4},
\end{aligned}
$$

(6.37)

with linearization constants:

$$k_{q1} = \frac{\partial K}{\partial c_{aj}(t)}(\bar{v}^*) = \frac{\partial K_a}{\partial c}(c_{aj}^*(t), h_s^*(t)) + \sum_{f_{in}} \frac{\partial K_f}{\partial c}(c_{aj}^*(t), Q_f^*(t))$$

$$+ \frac{\partial K_f}{\partial c}(c_{aj}^*(t), Q_u^*(t)),$$

$$k_{q2} = \frac{\partial K}{\partial h_s(t)}(\bar{v}^*) = \frac{\partial K_a}{\partial h_s}(c_{aj}^*(t), h_s^*(t)) - (1 - D_{rj}(t)\Delta t)\frac{\partial K_a}{\partial h_s}(c_{aj}^*(t-1), h_s^*(t)),$$

$$k_{q3} = \frac{\partial K}{\partial c_{aj}(t-1)}(\bar{v}^*) = -(1 - D_{rj}(t)\Delta t)\frac{\partial K_a}{\partial c}(c_{aj}^*(t-1), h_s^*(t)), \qquad (6.38)$$

$$k_{qf1} = \frac{\partial K}{\partial c_{fin}(t)}(\bar{v}^*) = -\frac{\partial K_f}{\partial c}(c_{fj}^*(t), Q_f^*(t)), \quad \forall f_{in},$$

$$k_{qf2} = \frac{\partial K}{\partial Q_{fin}(t)}(\bar{v}^*) = -\frac{\partial K_f}{\partial Q}(c_{fj}^*(t), Q_f^*(t)) + \frac{\partial K_f}{\partial Q}(c_{aj}^*(t), Q_f^*(t)), \quad \forall f_{in},$$

$$k_{q4} = K(\bar{v}^*) - k_{q1}c_{aj}^*(t) - k_{q2}h_s^*(t) - k_{q3}c_{aj}^*(t-1) - \sum_{f_{in}}[k_{qf1}c_{fj}^*(t) + k_{qf2}Q_f^*(t)],$$

in which the following are forward-estimated parameters:

$\bar{v}^*$ : vector of forward-estimated optimization variables;

$c_{aj}^*$ : concentration of pollutant $j$ in the surface-water subsystem in the previous cycle (mg/l);

$h_s^*$ : surface-water level in the previous cycle (m+ref);

$c_{fj}^*$ : concentration of pollutant $j$ of flow element $f$ which delivers to the current surface-water subsystem in the previous cycle (mg/l);

$Q_f^*$ : discharge of flow element $f$ in the previous cycle (m³/s);

$Q_u^*$ : uncontrolled inflow into the surface-water subsystem in the previous cycle (m³/s).

For groundwater-flow elements, water-quality transport is only included if the pollutant is conservative, which means that its decay rate $D_r$ equals zero.


## 6.4   Flow-Element Modeling

Flow elements connect the areas in a water system and transport water and pollutants. Three groups of flow elements can be distinguished (Sec. 4.3.1):
regulating structures:
- pumping stations,
- weirs,
- sluices,
- inlets,

fixed structures:
- weirs,
- inlets,

free flow elements:
- canals,
- groundwater.

All flow elements distinguished, play a role in the continuity constraints of groundwater and surface-water subsystems of areas. Some flow elements impose additional side constraints to the optimization problem, in which the physical relationships describing flow are formulated i.e.: weirs, sluices, canals and groundwater.

### 6.4.1 Pumping Stations

Pumping stations often contain several pumping units. To keep the optimization model as small as possible, the choice has been made to model all pumping units together in one flow element in optimization and impose the availability of these units by means of lower and upper limits on the respective model variable.

To prevent unnecessary operation of a pumping station, the discharge variable $Q_p$ is included in the objective function (Eq 6.6), with a very small penalty coefficient $W_r$. This penalty is chosen so small that it does not interfere with the objective of maintaining the targets set for interests in the water system by means of damage functions.

Optimization yields the required discharge of a pumping station $Q_p$, which may be smaller or equal to the maximum available capacity of the pumping station. If the required discharge is smaller, it generally cannot be achieved by discrete capacities of pumping units. To deal with this problem and include the option of preferred operational sequences and minimal on- and off-periods for pumping units, a pump-selection algorithm has been developed. This algorithm, which is implemented in the Constraint Manager, is briefly discussed below and presented schematically in Fig. 6.16.

The pump-selection algorithm is used to determine the combination of available pumps that best meets the required discharge $Q_p$ of the pumping station at optimization time step $t = 2$ in each previous loop. This discharge is subsequently imposed on the optimization problem, by setting the lower and upper limits on variable $Q_p$ at optimization time $t = 1$ to $Q_{pi}$. This method guarantees a feasible discharge produced by optimization for time step $t = 1$, which can be used in simulation without modification (see also Fig. 6.1).

The algorithm features an operational threshold, which prevents a pumping unit from being switched on, if the required capacity is below the threshold value. In case the required capacity is low, this feature can postpone operation of the pumping station. In general, during subsequent loops, the required pumping capacity at $t = 2$ will increase, if switching on of a pump is postponed. A pump is only switched on once the threshold is passed. In other situations, a small required discharge could disappear. This situation can occur if the required discharge results from prediction errors that decrease during subsequent loops. The situation could also occur if excess water leaves a surface-water subsystem via other flow elements or by evaporation.

A special form of LP, called Mixed Integer Linear Programming (MIP), could circumvent some of the problems described. Using MIP, variables can be defined to have an integer value only. This feature enables pumping units of a pumping station to be switched independently, while optimization would automatically require a unit to be on or off. However, a MIP problem thus defined would require each pump to be represented by a separate variable, each of which would have to occur in the objective function with a separate penalty to force the operational sequence. This would increase the problem considerably in size and lead to a reduction in solution speed. Therefore MIP has not been considered here.

In case of electrical pumping stations, in addition to the operational restrictions described above, it may be necessary to prevent operation of pumps during peak hours. This is included in the optimization problem by setting the upper and lower limits on flow variable $Q_p$ to zero during those periods.

Furthermore, for electrical pumping stations, different tariffs for electricity may apply during the day and at night. Use of night tariffs is generally advantageous, since it reduces the running cost of the pumping station. The preference for using night hours in optimization, is arranged by reducing the penalty for operation of the pumping stations for these periods in the objective function.

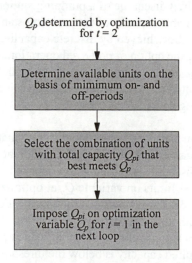

Fig. 6.16. Schematic pump-selection algorithm.

## 6.4.2  Weirs

Weir flow elements can be either controlled or fixed. In general, a well-designed fixed weir allows flow in one direction only. Controllable weirs have fixed upper and lower limits on the crest level. If the water level to be controlled exceeds this limit, the controllable weir behaves like a fixed weir, until the water level enters the range of adjustment again.

In general, backflow via a weir is only possible under exceptional circumstances. The specific situation of backflow via a weir is not included in optimization. However, if it should be modeled, two weirs can be included in the model, with opposite flow directions.

If the downstream water level is below the crest level, the flow condition of the weir is considered unsubmerged, whereas if it is above the crest level, the weir is considered to be submerged. In the submerged situation, both the upstream and the downstream water level influence the flow via the weir (Sec. 4.3.3). The general properties of the weir considered in optimization are shown in Fig. 6.17.

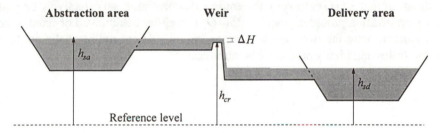

*Fig. 6.17. Properties of a weir flow element as modeled in optimization.*

To describe both submerged and unsubmerged flow over a rectangular weir, the following equation and corresponding restrictions have been introduced in Sec. 4.3.3:

$$Q_w = \begin{cases} C_w \, w_w \, (\Delta H + \frac{3}{2} H_1) \, \Delta H^{\frac{1}{2}} & , \ \Delta H > 0, \\ 0 & , \ \Delta H \le 0, \end{cases} \tag{6.39}$$

$$\begin{aligned} \Delta H &= H_2 - H_1, \\ H_1 &= h_{sd} - h_{cr}, \\ H_2 &= h_{sa} - h_{cr}, \end{aligned} \tag{6.40}$$

in which:
$Q_w$    : discharge over a weir (m³/s);
$C_w$    : weir discharge coefficient (m^½/s);
$w_w$    : weir-crest width (m);

$H_1$     : piezometric head to the crest downstream the weir (m);
$H_2$     : piezometric head to the crest upstream the weir (m);
$\Delta H$     : water-level difference (m);
$h_{sa}$     : water level in the abstraction area (m+ref);
$h_{sd}$     : water level in the delivery area (m+ref);
$h_{cr}$     : weir-crest level (m+ref).

The possible flow conditions yield the following:
*     unsubmerged weir: $\Delta H = H_2$ and $H_1 = 0$;
*     submerged weir: $\Delta H = H_2 - H_1$.

The appropriate situation with respect to backflow and flow condition for a specific time step of the control horizon, can be determined by means of forward estimating. To prevent backflow, the binary parameter $\kappa_w$ is introduced:

$$\kappa_w = \begin{cases} 1, & \text{if } h_{sa} > h_{hd} \text{ and } h_{sa} > h_{cr}, \\ 0, & \text{otherwise}. \end{cases} \tag{6.41}$$

In the situation that $\kappa_w$ is set to zero, the entire side constraint can effectively be removed from the optimization problem, whereas flow $Q_w$ can either be removed or fixed to zero.

To incorporate the two flow conditions in one side constraint of the optimization model, the following binary parameter is introduced:

$$\kappa_{ws} = \begin{cases} 1, & \text{if } h_{sd} > h_{cr}, \\ 0, & \text{if } h_{sd} \le h_{cr}. \end{cases} \tag{6.42}$$

With this parameter, $\Delta H$ can be rewritten as:

$$\Delta H = H_2 - \kappa_{ws} H_1. \tag{6.43}$$

This can be substituted in the weir equation Eq. 6.39, which yields a nonlinear equation, which can be linearized using a first-order Taylor approximation (Eq. 6.4):

$$\begin{aligned} Q_w &= \kappa_w C_w w_w (H_2 - \kappa_{ws} H_1 + \tfrac{3}{2}\kappa_{ws} H_1)(H_2 - \kappa_{ws} H_1)^{\frac{1}{2}} \\ &= \kappa_w C_w w_w (H_2 + \tfrac{1}{2}\kappa_{ws} H_1)(H_2 - \kappa_{ws} H_1)^{\frac{1}{2}} \\ &= \kappa_w C_w w_w [h_{sa} - h_{cr} + \tfrac{1}{2}\kappa_{ws}(h_{sd} - h_{cr})][h_{sa} - h_{cr} - \kappa_{ws}(h_{sd} - h_{cr})]^{\frac{1}{2}} \\ &= \kappa_w C_w w_w [h_{sa} + \tfrac{1}{2}\kappa_{ws} h_{sd} - (1 + \tfrac{1}{2}\kappa_{ws}) h_{cr}][h_{sa} - \kappa_{ws} h_{sd} - (1 - \kappa_{ws}) h_{cr}]^{\frac{1}{2}} \\ &= Q_w(h_{sa}, h_{sd}, h_{cr}) \approx k_{w1} h_{sa} + k_{w2} h_{sd} + k_{w3} h_{cr} + k_{w4}, \end{aligned} \tag{6.44}$$

in which the following linearization constants represent:

$$k_{w1} = \frac{\partial Q_w}{\partial h_{sa}}(h_{sa}^*, h_{sd}^*, h_{cr}^*) = \kappa_w C_w w_w \left\{ [h_{sa}^* - \kappa_{ws} h_{sd}^* - (1 - \kappa_{ws}) h_{cr}^*]^{\frac{1}{2}} \right\}$$

$$+ \frac{1}{2} \kappa_w C_w w_w \left\{ [h_{sa}^* + \frac{1}{2}\kappa_{ws} h_{sd}^* - (1 + \frac{1}{2}\kappa_{ws}) h_{cr}^*][h_{sa}^* - \kappa_{ws} h_{sd}^* - (1 - \kappa_{ws}) h_{cr}^*]^{-\frac{1}{2}} \right.$$

$$k_{w2} = \frac{\partial Q_w}{\partial h_{sd}}(h_{sa}^*, h_{sd}^*, h_{cr}^*) = \kappa_w C_w w_w \left\{ \frac{1}{2}\kappa_{ws} [h_{sa}^* - \kappa_{ws} h_{sd}^* - (1 - \kappa_{ws}) h_{cr}^*]^{\frac{1}{2}} \right\}$$

$$- \frac{1}{2} \kappa_w C_w w_w \left\{ \kappa_{ws} [h_{sa}^* + \frac{1}{2}\kappa_{ws} h_{sd}^* - (1 + \frac{1}{2}\kappa_{ws}) h_{cr}^*][h_{sa}^* - \kappa_{ws} h_{sd}^* - (1 - \kappa_{ws}) h_{cr}^*]^{-\frac{1}{2}} \right\} \quad (6.45)$$

$$k_{w3} = \frac{\partial Q_w}{\partial h_{cr}}(h_{sa}^*, h_{sd}^*, h_{cr}^*) = \kappa_w C_w w_w \left\{ -(1 + \frac{1}{2}\kappa_{ws}) [h_{sa}^* - \kappa_{ws} h_{sd}^* - (1 - \kappa_{ws}) h_{cr}^*]^{\frac{1}{2}} \right\}$$

$$- \frac{1}{2} \kappa_w C_w w_w \left\{ (1 - \kappa_{ws}) [h_{sa}^* + \frac{1}{2}\kappa_{ws} h_{sd}^* - (1 + \frac{1}{2}\kappa_{ws}) h_{cr}^*][h_{sa}^* - \kappa_{ws} h_{sd}^* - (1 - \kappa_{ws}) h_{cr}^*]^{-\frac{1}{2}} \right.$$

$$k_{w4} = Q_w(h_{sa}^*, h_{sd}^*, h_{cr}^*) - k_{w1} h_{sa}^* - k_{w2} h_{sd}^* - k_{w3} h_{cr}^*,$$

in which the following are:
optimization variables:

$Q_w$   : flow over a weir (m3/s), $Q_w \in (-\infty, \infty)$;
$h_{sa}$   : surface-water level in abstraction area (m+ref), $h_{sa} \in (-\infty, \infty)$;
$h_{sd}$   : surface-water level in delivery area (m+ref), $h_{sd} \in (-\infty, \infty)$;
$h_{cr}$   : weir-crest level (m+ref), $h_{cr} \in (-\infty, \infty)$;

forward-estimated parameters:

$h_{sa}^*$   : surface-water level in abstraction area in the previous cycle (m+ref);
$h_{sd}^*$   : surface-water level in abstraction area in the previous cycle (m+ref);
$h_{cr}^*$   : weir-crest level in the previous cycle (m+ref);
$\kappa_w$   : binary weir parameter to prevent back flow (-), $\kappa_w \in \{0, 1\}$;
$\kappa_{ws}$   : binary weir parameter for submerged flow (-), $\kappa_{ws} \in \{0, 1\}$.

In optimization, the general restriction is included that the crest of both controlled and fixed weirs is restricted by the adjustable limits:

$$h_{cr\,min} \leq h_{cr} \leq h_{cr\,max}, \qquad (6.46)$$

in which:
$h_{cr\,min}$   : minimum weir-crest level (m+ref);
$h_{cr\,max}$   : maximum weir-crest level (m+ref).

In case of a fixed weir, $h_{cr}$ is kept fixed by means of setting both the maximum and minimum crest level to the appropriate crest-level of the weir, in fact preventing it from moving.

In case of a controlled weir, the limits on the optimization variable $h_{cr}$ can be further restricted by means of forward estimating, to guarantee gradual changes in the crest-level height during the control horizon.

Simulation follows the control strategy determined by optimization, controlling either crest level or flow. The latter is preferred in the practical situation that the weir is controlled by means of a local PI controller (Sec. 4.3.3) since this controller controls the flow more accurately, guaranteeing a gentle change in both the flow and the crest-level height.

### 6.4.3 Sluices

The sluices involved in optimization are primarily spill sluices that discharge to a boundary, e.g. the sea or a river with a water level which is predicted for the control horizon (Fig. 6.18). However, a sluice can also be used in a water-system model as a regulating structure between surface-water subsystems of different areas.

Sluices can have several gates which can be opened and closed separately on the basis of a defined preferred opening sequence.

The general discharge equation for a sluice with one or more sliding gates reads (Sec. 4.3.4):

$$Q_{sl} = \mu_{sl} w_{sl} H_{sl} \sqrt{2g} \, \Delta H^{1/2}, \tag{6.47}$$

in which:

$Q_{sl}$    : sluice discharge (m³/s);
$\mu_{sl}$    : contraction coefficient (-);
$w_{sl}$    : width of one sliding gate (m);
$H_{sl}$    : total opening of all gates together $= A_{sl} / w_{sl}$ (m);
$\Delta H$    : difference in water level between abstraction and delivery area (m);
$g$    : gravitational constant (m/s²).

This equation assumes only the sliding gates restrict the flow through the sluice. Furthermore, submerged flow under equally sized sliding gates is assumed and a contraction coefficient $\mu_{sl}$ which is independent of the number of gates opened and the opening height.

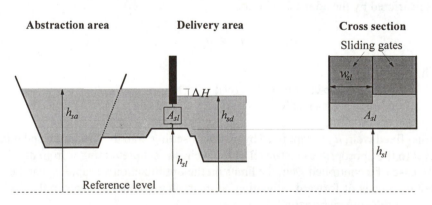

*Fig. 6.18. Properties of a sluice flow element as modeled in optimization.*

In optimization a binary parameter $\kappa_{sl}$ is used to prevent backflow through the sluice:

$$\kappa_{sl} = \begin{cases} 1 & \text{if } h_{sa} > h_{sd} \\ 0 & \text{otherwise}. \end{cases} \tag{6.48}$$

The nonlinear sluice equation can be linearized. Assuming the flow through the sluice being abstracted from an area and delivered to another area, the following side constraint to the optimization model is obtained:

$$
\begin{aligned}
Q_{sl} &= \kappa_{sl}\, \mu_{sl}\, w_{sl}\, H_{sl} \sqrt{2g}\, \Delta H^{\frac{1}{2}} = \kappa_{sl}\, \mu_{sl}\, w_{sl} \sqrt{2g}\, H_{sl} \left( h_{sa} - h_{sd} \right)^{\frac{1}{2}} \\
&= Q_{sl}(h_{sa}, h_{sd}, H_{sl}) \approx k_{sl1}\, h_{sa} + k_{sl2}\, h_{sd} + k_{sl3}\, H_{sl} + k_{sl4},
\end{aligned}
\tag{6.49}
$$

in which the following linearization constants represent:

$$
\begin{aligned}
k_{sl1} &= \frac{\partial Q_{sl}}{\partial h_{sa}} (h_{sa}^*, h_{sd}^*, H_{sl}^*) = \tfrac{1}{2}\, \kappa_{sl}\, \mu_{sl}\, w_{sl} \sqrt{2g}\, H_{sl}^* \left( h_{sa}^* - h_{sd}^* \right)^{-\frac{1}{2}}, \\
k_{sl2} &= \frac{\partial Q_{sl}}{\partial h_{sd}} (h_{sa}^*, h_{sd}^*, H_{sl}^*) = -\tfrac{1}{2}\, \kappa_{sl}\, \mu_{sl}\, w_{sl} \sqrt{2g}\, H_{sl}^* \left( h_{sa}^* - h_{sd}^* \right)^{-\frac{1}{2}}, \\
k_{sl3} &= \frac{\partial Q_{sl}}{\partial H_{sl}} (h_{sa}^*, h_{sd}^*, H_{sl}^*) = \kappa_{sl}\, \mu_{sl}\, w_{sl} \sqrt{2g} \left( h_{sa}^* - h_{sd}^* \right)^{\frac{1}{2}}, \\
k_{sl4} &= Q_{sl}(h_{sa}^*, h_{sd}^*, H_{sl}^*) - k_{sl1}\, h_{sa}^* - k_{sl2}\, h_{sd}^* - k_{sl3}\, H_{sl}^*,
\end{aligned}
\tag{6.50}
$$

in which the following are:
optimization variables:
$Q_{sl}$    : sluice discharge (m3/s), $Q_{sl} \in (-\infty, \infty)$;
$H_{sl}$    : total gate opening (m), $H_{sl} \in [0, \infty)$;
forward-estimated parameters:
$h_{sa}^*$    : surface-water level in abstraction area in the previous cycle (m+ref);
$h_{sd}^*$    : surface-water level in the delivery area in the previous cycle (m+ref);
$H_{sl}^*$    : total height of gate openings in the previous cycle (m);
$\kappa_{sl}$    : binary parameter to prevent backflow through a sluice (-).

In the linearization above, the sluice is assumed to discharge to a surface-water subsystem of an area. If the sluice discharges to a boundary, the linearization constant $k_{sl2}$ equals zero and $h_{sd}$ is made available to optimization by the prediction model via time-series data.

The discharge of the sluice and the total gate opening are the only variables taken into account in the optimization model. The optimization result that is used by simulation, is the total gate opening height. In simulation, the total gate opening height is distributed over the available number of sliding gates, according to the preferred opening and closing sequence.

Furthermore, a minimum opening threshold is included in simulation to prevent the sliding gates from having a very small opening.

In practice, small errors resulting from the assumptions about a constant contraction coefficient, made for the sluice discharge equation Eq. 6.47, are corrected automatically in each subsequent loop. When the discharge calculated in optimization is too large, an increased fall in water level in the abstraction area is measured. Consequently, the next optimization stipulates to open the sliding gates less far. This corrective mechanism only applies if the errors are relatively small compared to the total discharge.

### 6.4.4 Inlets

Inlets are considered to have a fixed maximum capacity which is independent of the water levels up- and downstream. To prevent backflow, the binary parameter $\kappa_{il}$ is included in optimization:

$$\kappa_{il} = \begin{cases} 1 & \text{if } h_{sa} > h_{sd}, \\ 0 & \text{otherwise}. \end{cases} \tag{6.51}$$

In optimization, only the discharge variable $Q_{il}$ is included, which is restricted between an operational threshold and the maximum inlet capacity.

To prevent unnecessary operation, the discharge variable $Q_{il}$ is included in the objective function (Eq 6.6) with only a very small penalty coefficient $W_r$, as in modeling of a pumping station.

### 6.4.5 Canals

Canals form a special category of flow elements, since they allow flow in two directions (Fig. 6.19). This affects water-quantity and water-quality modeling. As mentioned in Sec. 4.3.6, constant flow through a canal, under specific assumptions, can be described by the Chézy equation. In this equation, the hydraulic gradient $i_c$ determines the direction of flow and can be negative. However, the Chézy equation is not suitable for determining negative flow. Therefore, the reversion factor $\gamma_c$ is used:

$$\gamma_c = \begin{cases} 1 & \text{if } i_c \geq 0, \\ -1 & \text{if } i_c < 0. \end{cases} \tag{6.52}$$

Thus the flow-velocity equation becomes:

$$v_c = \gamma_c C_c (\gamma_c R i_c)^{\frac{1}{2}},$$
$$C_c = 18 \, {}^{10}\log\left(\frac{12 R}{k_N}\right), \tag{6.53}$$

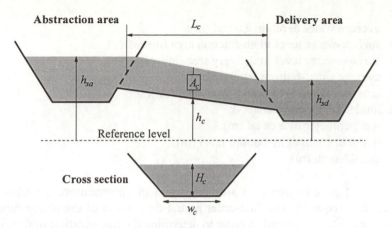

*Fig. 6.19. Properties of a canal flow element as modeled in optimization.*

in which:

$v_c$     : flow velocity in canal (m/s);
$\gamma_c$     : canal flow reversion factor (-);
$C_c$     : Chézy coefficient (m$^{\frac{1}{2}}$/s);
$R$     : hydraulic radius (m);
$i_c$     : hydraulic gradient (-);
$k_N$     : Nikuradse roughness coefficient (m).

For rural canal systems, the Chézy coefficient $C_c$ varies only very little. Therefore, the coefficient is assumed to be constant and equal to the coefficient corresponding to the first optimization time step, for the entire control horizon.

The Chézy equation depends on the water levels in the abstraction and the delivery areas as stated in the following equations:

$$
\begin{aligned}
H_c &= \tfrac{1}{2}\left(h_{sa} + h_{sd}\right) - h_c, \\
A_c &= \left(w_c + n_c H_c\right) H_c, \\
P_c &= w_c + 2\left(n_c^2 + 1\right)^{\frac{1}{2}} H_c, \\
R &= \frac{A_c}{P_c}, \\
i_c &= \frac{h_{sa} - h_{sd}}{L_c},
\end{aligned}
\tag{6.54}
$$

in which:

$H_c$ : average water depth in a canal (m);
$h_{sa}$ : surface-water level in abstraction area (m+ref);
$h_{sd}$ : surface-water level in delivery area (m+ref);
$h_c$ : average canal-bottom level (m+ref);
$A_c$ : cross-sectional area half-way a canal (m²);
$n_c$ : total side slope of canal (-);
$P_c$ : wet perimeter of a canal (m);
$w_c$ : width of canal at its bottom level (m);
$L_c$ : canal length (m).

To include the flow equation as a side constraint in optimization, a first-order Taylor approximation is required. The first-order partial derivatives of the above functions with respect to $h_{sa}$ and $h_{sd}$ are required in order to determine the linearization of the velocity side constraint:

$$
\begin{aligned}
v_c &= \gamma_c \, C_c \, (\gamma_c \, R \, i_c)^{\frac{1}{2}} \\
&= v_c (h_{sa}, h_{sd}) \approx k_{c1} \, h_{sa} + k_{c2} \, h_{sd} + k_{c3} \, .
\end{aligned}
\tag{6.55}
$$

The derivatives of the product $(\gamma_c \, R \, i_c)$ are needed for the linearization:

$$
\frac{\partial (\gamma_c R \, i_c)}{\partial h_{sa}} (h_{sa}^*, h_{sd}^*) = \gamma_c \left[ i_c (h_{sa}^*, h_{sd}^*) \frac{\partial R}{\partial h_{sa}} (h_{sa}^*, h_{sd}^*) + R (h_{sa}^*, h_{sd}^*) \frac{\partial i_c}{\partial h_{sa}} (h_{sa}^*, h_{sd}^*) \right],
$$

$$
\frac{\partial (\gamma_c R \, i_c)}{\partial h_{sd}} (h_{sa}^*, h_{sd}^*) = \gamma_c \left[ i_c (h_{sa}^*, h_{sd}^*) \frac{\partial R}{\partial h_{sd}} (h_{sa}^*, h_{sd}^*) + R (h_{sa}^*, h_{sd}^*) \frac{\partial i_c}{\partial h_{sd}} (h_{sa}^*, h_{sd}^*) \right].
\tag{6.56}
$$

The derivatives of the average canal depth $H_c$ are:

$$
\frac{\partial H_c}{\partial h_{sa}} (h_{sa}^*, h_{sd}^*) = \frac{\partial H_c}{\partial h_{sd}} (h_{sa}^*, h_{sd}^*) = \frac{1}{2} \, .
\tag{6.57}
$$

The derivatives of the cross-sectional area $A_c$ are:

$$
\frac{\partial A_c}{\partial h_{sa}} (h_{sa}^*, h_{sd}^*) = \frac{\partial A_c}{\partial h_{sd}} (h_{sa}^*, h_{sd}^*) = \frac{\partial A_c}{\partial H_c} \frac{\partial H_c}{\partial h_{sd}} (h_{sa}^*, h_{sd}^*) = \frac{1}{2} w_c + n_c H_c (h_{sa}^*, h_{sd}^*).
\tag{6.58}
$$

The derivatives of the wet perimeter $P_c$ are:

$$
\frac{\partial P_c}{\partial h_{sa}} (h_{sa}^*, h_{sd}^*) = \frac{\partial P_c}{\partial h_{sd}} (h_{sa}^*, h_{sd}^*) = \frac{\partial P_c}{\partial H_c} \frac{\partial H_c}{\partial h_{sd}} (h_{sa}^*, h_{sd}^*) = (n_c^2 + 1)^{\frac{1}{2}}.
\tag{6.59}
$$

The derivatives of the hydraulic radius $R$ are:

$$\frac{\partial R}{\partial h_{sa}}(h_{sa}^*, h_{sd}^*) = \frac{\partial R}{\partial h_{sd}}(h_{sa}^*, h_{sd}^*)$$

$$= \frac{\dfrac{\partial A_c}{\partial h_{sd}}(h_{sa}^*, h_{sd}^*)}{P_c(h_{sa}^*, h_{sd}^*)} - \frac{\dfrac{\partial P_c}{\partial h_{sd}}(h_{sa}^*, h_{sd}^*) A_c(h_{sa}^*, h_{sd}^*)}{P_c^2(h_{sa}^*, h_{sd}^*)} \qquad (6.60)$$

$$= \frac{\tfrac{1}{2}w_c + n_c H_c(h_{sa}^*, h_{sd}^*)}{P_c(h_{sa}^*, h_{sd}^*)} - \frac{(n_c^2 + 1)^{\frac{1}{2}} A_c(h_{sa}^*, h_{sd}^*)}{P_c^2(h_{sa}^*, h_{sd}^*)}.$$

The derivatives of the hydraulic gradient $i_c$ are:

$$\frac{\partial i_c}{\partial h_{sa}}(h_{sa}^*, h_{sd}^*) = -\frac{\partial i_c}{\partial h_{sd}}(h_{sa}^*, h_{sd}^*) = \frac{1}{L_c}. \qquad (6.61)$$

Finally, the linearization constants of Eq. 6.55 represent:

$$k_{c1} = \frac{\partial v_c}{\partial h_{sa}}(h_{sa}^*, h_{sd}^*) = \tfrac{1}{2}\gamma_c C_c \left[\gamma_c R(h_{sa}^*, h_{sd}^*)\, i_c(h_{sa}^*, h_{sd}^*)\right]^{-\frac{1}{2}} \frac{\partial(\gamma_c R\, i_c)}{\partial h_{sa}}(h_{sa}^*, h_{sd}^*),$$

$$k_{c2} = \frac{\partial v_c}{\partial h_{sa}}(h_{sa}^*, h_{sd}^*) = \tfrac{1}{2}\gamma_c C_c \left[\gamma_c R(h_{sa}^*, h_{sd}^*)\, i_c(h_{sa}^*, h_{sd}^*)\right]^{-\frac{1}{2}} \frac{\partial(\gamma_c R\, i_c)}{\partial h_{sd}}(h_{sa}^*, h_{sd}^*), \qquad (6.62)$$

$$k_{c3} = v_c(h_{sa}^*, h_{sd}^*) - k_{c1} h_{sa}^* - k_{c2} h_{sd}^*,$$

in which the following are:
optimization variable:
$v_c$        : flow velocity in canal (m/s);
forward-estimated parameters:
$h_{sa}^*$     : surface-water level in the abstraction area in the previous cycle (m+ref);
$h_{sd}^*$     : surface-water level in the delivery area in the previous cycle (m+ref).

A relatively small time step $\Delta t$ is required to produce stable flow results. This is to a large extent the result of the stationary situation which is assumed in the simplified canal equation used here.

### 6.4.6 Groundwater

Groundwater-flow elements are used to model groundwater flow between different areas. The interaction of groundwater and surface water within a single area is discussed in Sec. 6.3.1.

Groundwater-flow elements can be used to build entire groundwater networks and describe subsurface flow only. However, in the present study the interaction between groundwater flow and surface-water flow is of special interest. Groundwater-flow elements can connect the following subsystems of different areas:
- a groundwater subsystem with another groundwater subsystem;
- a groundwater subsystem with a surface-water subsystem (Fig. 6.20);
- a surface-water subsystem with another surface-water subsystem.

Similar to canals, groundwater flow is nondirectional. The modeled one-dimensional groundwater-flow equation of Darcy, described in Sec. 4.3.7, is linear and can thus be incorporated in optimization without linearization:

$$Q_{gf} = \frac{A_{gf}K_{gf}(h_a - h_d)}{L_{gf}},$$ (6.63)

in which the following are:
optimization variables:
$Q_{gf}$    : groundwater flow (m³/s), $Q_{gf} \in (-\infty, \infty)$;
$h_a$    : water level in the abstraction area (m+ref), $h_a \in (-\infty, \infty)$;
$h_d$    : water level in the delivery area (m+ref), $h_d \in (-\infty, \infty)$;
model parameters:
$A_{gf}$    : cross-sectional area of groundwater flow element (m²);
$K_{gf}$    : hydraulic soil conductivity for groundwater flow element (m/s);
$L_{gf}$    : length of groundwater flow element (m).

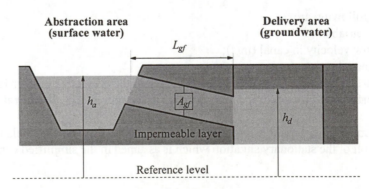

*Fig. 6.20. Properties of a groundwater-flow element as modeled in optimization.*

## 6.5 Mathematical Model Summary

This section presents a summary of the mathematical optimization problem. First, the overall objective function is given, followed by the continuity and side constraints. Each constraint is accompanied by references to the equation it is derived from. At the end of this section, Table 6.2 (Page 153) presents the modeling elements and the corresponding increase in size of the optimization problem.

Overall objective function:

$$
Z = \sum_{t=1}^{T} \left\{ \sum_{i=1}^{m_a} \left[ W_{ai} \sum_{j=1}^{n_i(t)} R_{di,j} \, \eta_{i,j}(t) + p_{us} \, Q_{us\,i}(t) \right] \right\}
$$

$$
+ \sum_{t=1}^{T} \left\{ \sum_{i=1}^{m_p} W_{p\,i}(t) \, Q_{p\,i}(t) + \sum_{i=1}^{m_{il}} W_{il\,i} \, Q_{il\,i}(t) \right\},
$$

(6.64)

in which:

| | | |
|---|---|---|
| $Z$ | : | objective function (-); |
| $t$ | : | index for optimization time steps, $t \in \{0, 1, .. , T\}$; |
| $T$ | : | control horizon in discrete time steps; |
| $i$ | : | index for areas, pumping stations and inlets; |
| $m_a$ | : | total number of areas; |
| $W_{ai}$ | : | weight of area $i$ (-), $W_a \in [0, 100]$; |
| $j$ | : | index for interests / damage functions; |
| $n_i(t)$ | : | total number of active damage functions in area $i$ at time step $t$; |
| $R_{di,j}$ | : | relative importance of interest $j$ in area $i$ (-), $R_d \in [0, 1]$; |
| $\eta_{i,j}(t)$ | : | damage variable of function $j$ defined for area $i$ at time step $t$ (-), $\eta \in [0, 1]$; |
| $p_{us}$ | : | large penalty coefficient for flow variable $Q_{us}$ (s/m$^3$); |
| $Q_{us\,i}$ | : | flow variable to prevent the surface-water level in area $i$ from falling below the bottom of the subsystem (m$^3$/s), $Q_{us} \in [0, \infty)$; |
| $m_p$ | : | total number of pumping stations; |
| $W_{p\,i}$ | : | penalty for flow via a pumping station $i$ (s/m$^3$), $W_p \in (0, \varepsilon]$; |
| $Q_{p\,i}$ | : | flow variable for pumping station $i$ (m$^3$/s); |
| $m_{il}$ | : | total number of inlets; |
| $W_{il\,i}$ | : | penalty for flow via an inlet $i$ (s/m$^3$), $W_{il} \in (0, \varepsilon]$; |
| $Q_{il\,i}$ | : | flow variable for inlet $i$, (m$^3$/s). |

### 6.5.1 Subsystems

*Continuity Constraints*

Groundwater subsystem, variables and parameters as given by Eq. 6.23:

$$h_g(t) = h_g(t-1) + \frac{1}{\mu^*}\left[-Q_g(t) + \sum Q_{gga}(t) + \frac{P_r^*(t) - C_r^*(t) + q_{ge}(t)}{3600 \cdot 1000}\right]\Delta t. \quad (6.65)$$

Surface-water subsystem, variables and parameters as given by Eq. 6.29 and 6.31:

$$k_{hs1}\,h_s(t) + k_{hs2}\,h_s(t-1) + k_{hs3} = Q_g(t) + \sum Q_f(t) + Q_{us}(t)$$
$$+ \frac{A_s\,q_{se}(t)}{3600 \cdot 1000} - \frac{A_s\,[P(t) - E_0(t)]}{1000\,\Delta t}. \quad (6.66)$$

Surface-water quality, variables and parameters as given by Eq. 6.38:

$$[k_{q1} + \sum_{f_{out}} k_{qf1}]\,c_{aj}(t) + k_{q2}\,h_s(t) + k_{q3}\,c_{aj}(t-1) + k_{q4}\,h_s(t-1)$$
$$- \sum_{f_{in}} [k_{qf1}\,c_{fj}(t) + k_{qf2}\,Q_f(t)] + \sum_{f_{out}} [k_{qf2}\,Q_f(t)] + k_{q5}\,Q_{us}(t) + k_{q6} = K_{uj}(t) \quad (6.67)$$

### *Side Constraints*

Damage functions should be defined piecewise linear and convex. For each line segment $j$ an $\eta$-inequality side constraint for variable $x$ should be defined, e.g. for water level $h$; for pollutant $c$. Variables and parameters are given by Eq. 6.16:

$$p_j\,x + d_j \le \eta. \quad (6.68)$$

Soil outflow, variables and parameters as given by Eq. 6.26:

$$Q_g = A_g\left[\left(\frac{1}{S_{rg}} + \frac{\kappa_d}{S_{rd}}\right)h_g - \left(\frac{1}{S_{rg}} + \frac{\kappa_s}{S_{rd}}\right)h_s + \frac{(\kappa_s - \kappa_d)}{S_{rd}}\,h_{dr}\right]. \quad (6.69)$$

### 6.5.2  Flow Elements

Weir side constraint, variables and parameters as given by Eq. 6.45:

$$Q_w = k_{w1}\,h_{sa} + k_{w2}\,h_{sd} + k_{w3}\,h_{cr} + k_{w4}. \quad (6.70)$$

Sluice side constraint, variables and parameters as given by Eq. 6.50:

$$Q_{sl} = k_{sl1} h_{sa} + k_{sl2} h_{sd} + k_{sl3} H_{sl} + k_{sl4}. \tag{6.71}$$

Canal side constraint, variables and parameters as given by Eq. 6.62:

$$v_c = k_{c1} h_{sa} + k_{c2} h_{sd} + k_{c3}. \tag{6.72}$$

Groundwater-flow side constraint, variables and parameters as given by Eq. 6.63:

$$Q_{gf} = \frac{A_{gf} k_{gf} (h_a - h_d)}{L_{gf}}. \tag{6.73}$$

Table 6.2. Modeling elements that determine the size of the optimization problem.

| Modeling element | | Number of variables | Variable name | Number of equations | Type of equation |
|---|---|---|---|---|---|
| Boundary | | 0 | - | 0 | - |
| Area | Surface water | | | | |
| | • Quantity | 2 | $h_s$, $Q_{us}$ | 1 | Continuity |
| | • Quality (for each variable $j$) | 1 | $c_j$ | 1 | Continuity |
| | Groundwater | | | | |
| | • Quantity | 1 | $h_g$ | 1 | Continuity |
| | • In-/outflow | 1 | $Q_g$ | 1 | Side constraint |
| Pumping station | | 1 | $Q_p$ | 0 | - |
| Weir | | 2 | $Q_w$, $h_{cr}$ | 1 | Side constraint |
| Sluice | | 2 | $Q_{sl}$, $H_{sl}$ | 1 | Side constraint |
| Inlet | | 1 | $Q_{il}$ | 0 | - |
| Groundwater flow | | 1 | $Q_{gf}$ | 1 | Side constraint |
| Damage functions | Each function | 1 | | 0 | - |
| | Each line segment | 0 | | 1 | Constraint |

## 6.6   Concluding Remarks

In this chapter an optimization model to control a general water system has been described. Special attention has been paid to modeling approaches. The general LP approach was chosen because it is flexible, relatively easy to implement and maintain, and can be scaled well. The details of the latter are outside the scope of the current study, but certainly of importance when considering the speed with which the optimal solution can be found by a solver. In a well-scaled optimization problem the optimal values of variables are all of the same order of magnitude, which reduces the chance of numerical instability during optimization and prevents long iteration times.

The solution speed is furthermore dependent on the solver used. In the present study the Interior Point method is chosen for solving practical optimization problems. This method has proven to be the most suitable for the general LP problem described here.

For the various types of structures which can be distinguished in a water system, the equations that should be incorporated in the optimization problem have been derived. Nonlinear relationships have been linearized using first-order Taylor approximations.

In Chapter 7, the general optimization problem described here is applied in several practical cases. Furthermore, at the end of that chapter, considerations with respect to the speed of the solving process will be given.

# 7  Case Studies

## 7.1  Introduction

### 7.1.1  Main Objectives

This chapter presents the application of the methods developed and described in the preceding chapters. All calculations have been made using the Decision Support System AQUARIUS, which has been developed as part of the present research.

The selected case studies are representative for the majority of regional water systems in the Netherlands. The locations of the areas described are indicated on the location map of Fig. 1.1. From a survey among 11 water boards (Lobbrecht, 1994c), five practical case studies have been selected and carried out in cooperation with other researchers, who did their work within the framework of the present research. Three of these studies are described in this chapter (Delfland, De Drie Ambachten and Salland). Some interesting results of the two others (Fleverwaard and Mark en Weerijs), are only briefly discussed in the concluding remarks of this chapter.

The first case study describes the practical situation in the Delfland area, which includes a large number of polders that drain into a storage basin. Of special interest is the combined functioning of the polders and the storage basin, to satisfy the main interests in the area: flood-prevention, ecology, agriculture, recreation and navigation. Both water-quantity and water-quality aspects play a role in controlling the Delfland water system, depending on the season and the hydrological load.

The second case study comprises a water system which is part of the De Drie Ambachten area. This is a gently sloping water system which drains by gravity to a tidal river. Recreational interests prevail in and along the surface-water subsystems of this water system. Strictly maintaining the surface-water level is of utmost importance since it entirely determines whether the requirements of the recreational interests along the beaches of the creek are satisfied. The water system is controlled by water authorities of both Belgium and the Netherlands.

The third case study involves a hilly and sandy water system in the Salland area. In this area, the hydraulic conductivity of the soil is relatively high. As usual for this type of area, over-drainage by means of extensive canal systems has taken place. Water shortages occur as a result of low groundwater levels, especially in summer. For this area the combination of agricultural and nature interests requires water preservation and supply of alien water.

The objective of the case studies is to show and compare the results of various alternative control modes, with a main focus on dynamic control. The potential of dynamic control is assessed in various types of water systems, using time-series analysis. Within the scope of this thesis it is not possible to show all aspects of dynamic control. One of the problems inherent to describing a dynamic process is that the perception of movement, so essential in dynamic control, is lost in print.

The following modes of control are distinguished in the case studies (Sec. 2.4):
• Local Manual,
• Local Automatic,
• Dynamic Manual,
• Dynamic Automatic.

In most of the water systems studied, local-manual and local-automatic control modes are practiced. In modeling the water systems, these two modes are compared. Local manual control is assumed to be the regular mode in all models built. In comparison to local manual control, local automatic control generally presents an improvement in satisfying the requirements of interests in a water system. In all cases, further improvement can be achieved by dynamic manual and dynamic automatic control. Central control modes are not explicitly discussed in the case studies.

The case studies have been selected in such a way that the options of water-system control can be shown in various forms. Table 7.1 shows the main focus of the case studies.

Table 7.1. Main focus of the case studies.

| | Delfland | De Drie Ambachten | Salland |
|---|---|---|---|
| Rural subsystems | Yes | Yes | Yes |
| Urban subsystems | Yes | - | - |
| Water quantity control<br>  surface water<br>  groundwater | Yes<br>- | Yes<br>- | Yes<br>Yes |
| Water quality control | Yes | - | - |
| Prediction sensitivity analysis | Yes | Yes | - |

## 7.1.2  Damage Functions and Interest Weighing

Damage functions have been constructed for the various interests distinguished in the case studies. As described in Sec. 3.1, each interest is represented by one or more key variables of the water-system. Figure 7.1 shows examples of simple damage functions. As described before, these damage functions are used to steer the strategy determination process and therefore do not necessarily represent economical damage to interests.

For the functions in Fig. 7.1, the ranges in which damage is absent or very low, have been indicated by arrows. These ranges are here called *preferred ranges*, which indicates that the requirements of that interest are satisfied within the range. Outside the range, the requirements are not satisfied and the interest and the water system are considered to fail under these circumstances.

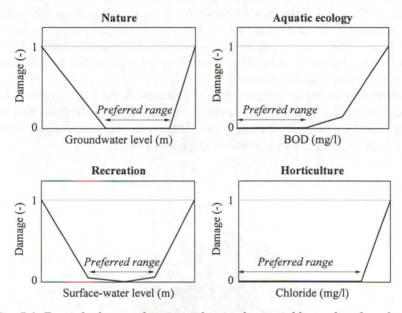

*Fig. 7.1. Example damage functions, showing key variables and preferred ranges.*

During low system loads, which may occur most of the time, local automatic and dynamic automatic control may produce similar control strategies. For some water systems it is therefore not always necessary to continuously use dynamic control. An option to switch over from local automatic to dynamic automatic control and vice versa, is therefore included. Choosing the right moment for switching from local to dynamic automatic control is essential to the success of dynamic control. Anticipating imminent high system-loads is one of the important features of dynamic control. This is why it is necessary to determine the right moment for switching very accurately.

Switching between local automatic and dynamic automatic control takes place in various examples of the cases described. The moment for switching is determined by simulating the water system in local automatic control mode during the control horizon. If during that control horizon the total damage to interests at the present simulation time step rises above a specified limit, dynamic control is automatically activated. The damage limit specified is called the *dynamic control threshold*.

The weighing of interests is visually represented in a chart, of which an example is shown in Fig. 7.2. The sizes of the inner sectors of the chart determine the relative weights assigned to various interests. The larger a sector, the more important the interest. The figure shows common-good interests and sectoral interests, which all have equal weights (see Table 1.1 on page 19, for example descriptions of the interests).

Interests are related to subsystems of the water system via the key variables, of which the state determines the satisfaction of the requirements according to the damage functions described above. These key variables are represented by the sectors of the hatched inner ring of the chart. The figure shows that one interest can be covered by several key variables, e.g.: groundwater level, surface water level, diffuse Biochemical Oxygen Demand (BOD) and the occurrence of alien substances. The number of variables per interest may vary and the same variable may be present in different interests.

The outer ring of the chart represents the locations in the water system where the various interests occur. Common-good interests such as flood prevention and ecology generally exist in all locations of the water system. Other interests, such as agriculture, water recreation and nature generally exist at one or more locations.

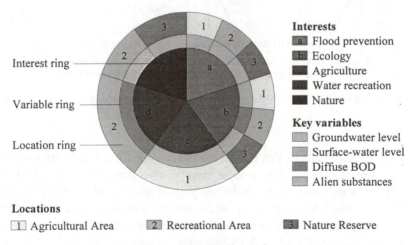

Fig. 7.2. The interest-weighing chart.

The choice of weights and shapes for the damage functions determine the optimum control strategy found in optimization. The ranges where damages are zero or only very low, are always preferred before any requirement of any interest in the water system is violated. This principle ensures that the capacity that is available in the water system is fully used before failure occurs. Performance of the water system can generally be improved simply by using this principle. As will be shown later, this means improvement of the average conditions required by interests and thus water-system performance.

If capacities available in the water system are not sufficient, the weights and damage functions determine the sequence in which interests in the water system will fail. In principle, the interest which has the lowest weight should fail first, followed by the interest with the second-lowest weight, etc. Using the highest weights for the most important, e.g. common-good interests, makes sure that these are the last to fail. If all damage functions in a water system are V-shaped, the sequence of failure can be determined easily. However, the shapes of most damage functions resembles parabola-shapes, which means that once the first interest has failed, a second interest may fail, while the first interest has not yet suffered the maximum damage. The weights and shapes of the curves thus together influence the extent of failure of interests under extreme conditions.

In general, the exact weighing of interests and the shapes of damage functions determine the optimal control strategy, especially under extreme conditions. In other situations, the control strategy is more or less independent of the weights and shapes of damage functions since the aim of the optimization is to prevent system failures. When all variables are within their preferred ranges, no significant improvement can be achieved by specifying the weights and shapes of damage functions more precisely.

The question arises which weights should be chosen for the various types of interests. In general, common-good interests will be allocated the highest weights. However, the sequence and extent of failure of sectoral interests is usually more difficult to determine. Special, multi-criteria techniques could be used to determine the required weights, but in the present research, another approach is used.

In this approach, frequency and duration of interest failures are decisive in the evaluation of the success of control under extreme conditions. In combination with time-series calculations, a particular set of weights results in a certain frequency and duration of failure. On the basis of evaluation, the decision makers of a water agency have to decide whether these properties are acceptable and alternatively, determine another set of weights. This approach emphasizes the system behavior required in control problems and reduces the need for determining weights formally, prior to implementing them. Parameters of frequency and duration of failures can be selected more easily by decision makers than weights. This prevents the users of a DSS from having to chose weights, the impact of which is uncertain or hard to determine in advance.

In the approach used, the selection of weights is an iterative and preferably interactive process, in which decision makers have to decide the required water-system behavior, rather than which standards to fix. This enables integrated decision making, considering the water system in its entirety, incorporating all interests. It furthermore requires decision makers to motivate their individual choices.

This approach has been used in the case studies, however, it was not used interactively involving all responsible decision makers of all water agencies. Such an exercise was considered outside the scope of the present study.

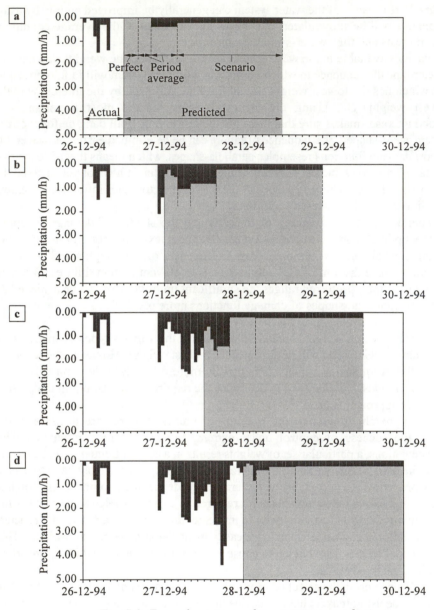

*Fig. 7.3. Example pictures of precipitation prediction.*

## 7.1.3 Hydrological-Load Prediction

As indicated in Table 7.1, the sensitivity of the prediction accuracy to the results of dynamic control is further analyzed in some practical case studies. Figure 7.3 shows how an excessive precipitation event of 50 mm could be predicted, using the methods discussed in Chapter 5.

The shaded frames show the prediction results for a control horizon of two days, as would be used in a time-series calculation at the beginning of each loop of Strategy Resolver. The following prediction scheme has been used (see also Table 5.3, Page 98):

- 1 - 4 hour: perfect prediction, uncertainty multiplier 0.6;
- 4 - 8 hour: period-average prediction, uncertainty multiplier 0.5;
- 8 - 16 hour: period-average prediction, uncertainty multiplier 0.5;
- 16 - 48 hour: wet scenario prediction.

In general, time steps of one hour are used in the models. Note that the pictures shown in Fig. 7.3 are only four of a total of 48 successive predictions required to calculate a time series of two days.

Uncertainty multipliers are used in the case studies to incorporate the general underestimates of excessive hydrological loads by the weather bureaus. In general, the most unfavorable situations have been used in the prediction. For example, an area with rapid runoff characteristics has been calculated using an uncertainty multiplier smaller than unity in the precipitation prediction.

### 7.1.4  Calibration and Accuracy

The various water systems modeled include the discharge and flow processes described in Chapter 4. Entire water systems have been modeled in detail for calibration purposes. In some cases the detailed models have been converted into *generalized models*. A generalized model is defined as a model of a water system which descries the hydrological processes that are important at a general scale, but is still accurate enough to represent the water system. The generalization is not a prerequisite, but it results in smaller optimization problems that can be solved efficiently.

It should be noted that in the term 'generalized model' used in this chapter, the word 'generalized' has a different meaning than in 'generalized network' as introduced in Chapter 3. The term generalized network is used in literature specifically for a special extension of network programming, which can be used for solving particular flow problems.

All detailed and generalized models have been calibrated against monitoring data available, following the approach shown in Fig. 7.4. To keep the models deterministic, model parameters have to be determined realistically and within normal ranges during calibration.

All models of the practical case studies presented have been calibrated by means of general water balances and data sets on flows and water levels. Monitoring data or sometimes only a few measurements have been used to make the models representative of the real water systems. In some cases, model parameters had to be used that were determined for other, similar water systems. The purpose has been to accurately describe the water-management situation and the extremes in system behavior.

The models built have been verified by means of verification data sets. In the descriptions, some, but certainly not all calibration results are presented.

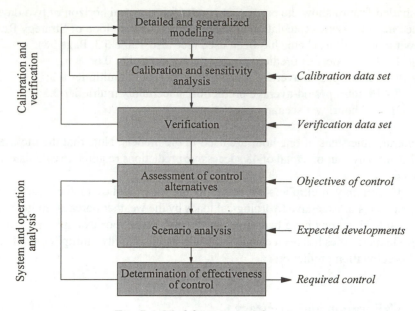

*Fig. 7.4. Modeling approach.*

Differences between data monitored and calculated have been observed in the calibration and verification processes. In general, these differences can have the following causes:
*   modeling inaccuracies,
*   measuring inaccuracies.

Models are reproductions of reality. For analysis purposes, it is of importance that a model closely resembles the general water-system behavior. It should be considered however, that exact reproductions cannot be obtained by this type of general model. This would require very detailed modeling, including each and every canal, structure, etc. Within the scope of the present study such detailed modeling is pointless and moreover, would have required detailed monitoring data which are generally not available.

   Inaccuracies in modeling can further result from incorrect initial system conditions. This subject has been studied thoroughly and resulted in the practice of using long initial runs for the majority of calculations described. This may result in an event of two days studied, which is preceded by an initial run of an entire year, to determine accurate system conditions at the start of the event.

In general, the data used in the models were not collected specifically for use in water-system modeling. Therefore, accuracy and resolution are generally not up to the required standard. For example, water-quality measurements taken twice a month, could be available from one location in a polder, whereas for an accurate calibration, daily measurements are needed at several locations. A recording error in such a single measurement, can have disastrous effects on the accuracy of the model calibrated.

The calibration data necessary, can generally be obtained for only very few locations in the water system. Precipitation and evaporation monitoring generally takes place outside the water-system area. Local precipitation effects have a major impact on runoff. Local precipitation as measured in rain gages may only result in local runoff, while in a model it could be assumed that the same precipitation fell in the entire water system. This effect can be limited in large water systems by using more than one measurement. However, geographically distributed data that have the required time resolution, are often not available.

Control alternatives have been formulated and assessed, on the basis of objectives of control set by the water authorities responsible for water management in the water systems studied (lower part of Fig. 7.4). Extra alternatives have sometimes been added, which were of interest from a research point of view.

Expected future developments in the water systems have been combined in development scenarios. These scenarios have been calculated, using various present and future control alternatives for regulating structures. On the basis of these results, the most advantageous modes of control have been determined.

### 7.1.5  Assessing Dynamic Control

Several parameters are used to evaluate the effectiveness of control modes, the most important one is the *performance index*. The performance index is used to determine the relative performance of water-system control for long periods, using a single control mode or a set of control modes.

The performance index is defined as the total sum of damages in a water system over time in a reference situation, divided by the total sum of damages when using an alternative set of control modes. The performance index is a general evaluation value, suitable to compare the effectiveness of control for all types of water systems. The performance index is determined by:

$$P_I \; = \; \frac{\sum D_{ref}}{\sum D_{alt}} ,$$

(7.1)

in which:

$P_I$        : performance index for an alternative control strategy (-);
$\sum D_{ref}$   : sum of water-system damages in a reference situation (-);
$\sum D_{alt}$   : sum of water-system damages resulting from an alternative set of control modes (-).

It should be realized that dynamic control as introduced in this thesis determines control strategies for periods of hours to days. To evaluate whether the requirements of some interests are satisfied, longer periods may sometimes be needed, e.g. the total duration a requirement is not satisfied in an entire year. Such data cannot be incorporated easily into the control strategy, simply because the control horizons used are too short.

In general, dynamic control will try to satisfy the long-term requirements of interests. If long-term effects should be incorporated, a long control horizon, e.g. several months, is needed. This approach can be followed to determine the required water system conditions for long periods, but not for day-to-day water control. Therefore, an approach is followed, in which indirect requirement definitions are used which are evaluated by the following parameters:

- the failure frequency;
- the failure duration;
- the flushing factor;
- the preservation factor.

The failure frequency is a variable used to assess the performance of the water system under extreme conditions, that lead to situations in which the requirements of one or more of the interests are violated.

As explained before, the requirements of interests are expressed in terms of water system variables, called key variables. The damage to an interest follows from the damage functions defined for that interest. If a key variable is outside the preferred range, than that requirement is considered to have been violated. When a defined combination of requirements fails, then the interest is considered to have suffered damage. If one or more interests have been damaged, the entire water system is considered to have failed. This situation is called a *system failure*.

The situations that cause a system failure can differ considerably and the effects of these failures can also differ much. For instance, failing to meet the requirements of a common-good interest, such as flood prevention, is generally considered more important than failing to meet the requirements of a sectoral interest, such as water recreation. However, this is not a prerequisite in the methodology developed.

The failure frequency is determined in time-series calculations, by recording the number of failures in the period calculated. The required calculations are carried out using the models built in the case studies.

In addition to the failure frequency, the duration of a system failure can be considered important. An example is the duration of a poor-water-quality situation in an urban area resulting from a combined sewer overflow.

With respect to water-quality control and satisfying the requirements of ecology-related interests, the quantity of alien water let into a water system or a subsystem can be important. To determine the quantity of water which is required for water-quality control, the *flushing factor* is introduced.

The flushing factor gives the ratio of water which is flushed through the water system each day for water-quality control, and the surface-water volume at the target level. The flushing factor is used to assess the quantity of alien let in for water-quality control as a result of various control options. If the flushing factor is zero, the system is not flushed for water quality-control:

$$F_F = \frac{\sum Q_{in\,quality}}{V_{st}}, \tag{7.2}$$

in which:

$F_F$      : flushing factor of the surface-water subsystem ($d^{-1}$);
$Q_{in\,quality}$ : inflow to maintain water quality ($m^3/d$);
$V_{st}$      : volume of surface-water subsystem at target water level ($m^3$).

Water preservation in hilly areas is considered an important issue for ecological interests. Ecological interests generally benefit from keeping as much location-specific water in a subsystem as possible, while preservation of water also minimizes the quantity which has to be supplied to a subsystem in times of drought.

To assess the effect of preserving water, the *preservation factor* is introduced, which determines the success of keeping as much water in a water system as possible over a particular period. The preservation factor is defined as the ratio of the quantity of water used and the total quantity abstracted from the system. In this respect the abstracted quantity is the sum of: water flowing out of the water system (e.g. by flow elements) with addition of the quantity used by the interests in the water system (e.g. total evapotranspiration).

$$P_F = \frac{\sum Q_{used}}{\sum Q_{out} + \sum Q_{used}}, \tag{7.3}$$

in which:

$P_F$      : preservation factor (-);
$Q_{used}$   : quantity of water used by interests ($m^3$);
$Q_{out}$    : outflow from a water system ($m^3$).


## 7.1.6  Forward Estimating Example

Forward estimating has been introduced in Sec. 6.2.2 as a method to overcome the disadvantage of linearizing nonlinear water-system behavior in optimization. The key feature in forward estimating is the iterative use of the previous optimization results in each forward-estimating cycle according to Fig. 6.2 (Page 112).

In the following, forward estimating is demonstrated by means of a time-series discharge calculation in a simple example water system. The example water system consists of an Area and a lower-lying Boundary, which are connected by two fixed parallel weirs (Fig. 7.5). As described in Sections 6.3.2 and 6.4.2, the relationship between water-level and volume in the surface-water subsystem, and the weir-discharge equation are both nonlinear.

The water level in the Area is always higher than that of the Boundary. The weirs have different crest-levels. The lowest weir is called 'Weir L', the other 'Weir H'. In the initial situation, the crest of Weir L is below the surface-water level of the Area and consequently it discharges from the Area into the Boundary. The crest of Weir H is above the initial surface-water level. During a high hydrological load, the surface-water level of the Area rises above the crest of Weir H, which should then also discharge.

Note that no regulating structures are incorporated in the example-water system. Therefore, the linearized problem formulated, only contains continuity and side constraints. This is done on purpose, to show the linearization results of successive forward-estimating cycles. In principle, the problem could be solved by means of elimination.

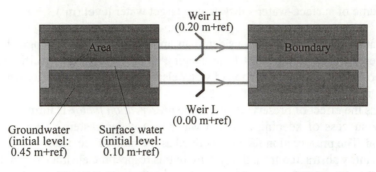

*Fig. 7.5. Example water system with two parallel weirs.*

Figure 7.6 shows four graphs which contain shaded frames for prediction and optimization results, with a time step $\Delta t$ of two hours, for a control horizon duration of two days. Graph (a) shows an extreme precipitation load of 50 mm. For simplicity, a perfect prediction of this hydrological load is assumed with an uncertainty multiplier of unity.

In the initial situation, no data on the development of the surface-water level in the Area are known. Therefore, in the first cycle (simulation time step $\tau = 0$), the initial surface-water level of 0.10 m+ref is used in linearizing the surface-water and weir equations (Eq. 6.29 and Eq. 6.44) for the entire control horizon.

Graph (b) shows that flow is determined by the optimization module for Weir L during the entire control horizon on the basis of the initial water-level. No flow can be determined for Weir H. These results are obtained by linearizing the weir equation for Weir L with an overflow height of 0.10 m for the entire control horizon. Weir H is excluded from the optimization problem in this cycle, because on the basis of the information available prior to the first cycle, it cannot discharge (Eq. 6.41 and 6.44). The result of the first cycle shows a clear water-level rise above the crest of Weir H during several time steps of the control horizon. This information is used in the second cycle.

In the second forward-estimating cycle, which represents the time two hours later (simulation time step $\tau = 1$), the water levels calculated in the first cycle are used as the basis for the linearizations. For both weirs, the flow equations are determined for each time step of the control horizon. Again, if the water level in the Area was predicted to be below the crest of Weir H in the previous cycle, the corresponding equation is excluded from the optimization problem, otherwise it is included. The result of the optimization module on the basis of these linearizations is shown in graph (c).

The last graph (d) shows the results which are obtained if the procedure described, is continued for two days. The results of simulation for these two days are shown in the left half of the graph. These results are obtained by using the full nonlinear surface water and weir equations and show that the estimates for water levels and flows over the two weirs

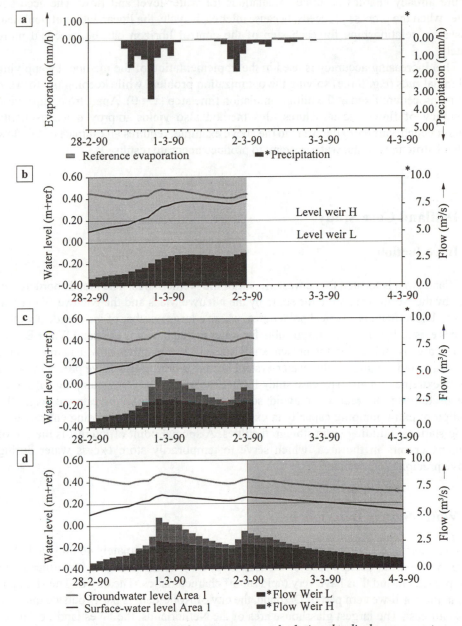

Fig. 7.6. Results of forward estimating in calculating the discharge over two parallel weirs for simulation time steps $\tau = 0$ (b), $\tau = 1$ (c) and $\tau = 24$ (d).

determined in the second forward-estimating cycle (graph c) using the linearized model, were fairly accurate already.

The procedure described, starts at simulation time $\tau = 0$, at which only the present data on water levels are available for forward estimating. After one cycle, the forward-estimating procedure already enables accurate calculations for water-level and flow. The accuracy improves when the process proceeds, because after each cycle, the linearization of nonlinear water-system relationships for each step of the control horizon can be executed more accurately.

This increasing accuracy is used in the implementation of the method, by applying several initial runs (e.g. three) solving the optimization problem while keeping the forward estimating procedure fixed at the initial simulation time step ($\tau = 0$). Apart from improving the accuracy of flows via structures, this method also yields improved water-quality calculation which, similar to the weirs in the above example, requires an estimate of the flow through all structures in the water system, to produce accurate results.

## 7.2   Delfland Case Study

### 7.2.1  Introduction

The Delfland Area is located in the western part of the Netherlands. The area is bordered in the west by the North Sea and in the south by the Nieuwe Maas and the Nieuwe Waterweg (see Fig. 1.1). Delfland is a typical polder area, where a large number of independent water-level areas exists. The authority responsible for water management is Delfland Water Board.

The area is included in the present study since it demonstrates how dynamic control can be used to reduce undesirable water-system behavior during excessive precipitation or periods of extreme drought. The case study further shows how existing pumping capacities can be used better to reduce or avoid scheduled future capital investment, e.g.: the installation of extra pumping capacity in existing pumping stations; building entirely new pumping stations; building extra storage facilities. A special point of interest is the use of retention reservoirs in the area, which serve to temporarily store excess water during excessive precipitation events.

### 7.2.2  Water-System Description

Delfland comprises 57 polders covering a total surface of approximately 40.000 ha. Urban and glasshouse areas occupy 35% of the area (Fig. 7.7). These areas are impermeable or semi-impermeable and thus have very rapid runoff characteristics. The city of The Hague is located in the north-western part of Delfland, the city of Rotterdam is partly outside the area, to the south-east. The largest glasshouse area of the Netherlands, the 'Westland', occupies the western part of the area.

The main surface-water subsystem is the Delfland Storage Basin, a system of interconnected water courses and lakes between the various polders, with a target water level of NAP - 0.40 m. Three types of area discharge into the storage basin: 'low-lying polders' (72%), 'high-lying polders' (6%) and 'storage-basin land' (22%). The water levels in the low-lying polders are below the water level of the storage basin; in the high-lying polders, the water levels are above that of the storage basin. The lowest target water level in the area is NAP - 6.30 m, whereas the highest is NAP + 1.40. Storage-basin land is the area along the storage basin that directly drains into the storage-basin surface water.

The main interests present in the area are: flood prevention, ecology, glasshouse horticulture, pasture agriculture (cattle farming) and navigation. These water-quantity and water-quality interests sometimes come into conflict.

Excess water is pumped from low-lying polders to the Delfland Storage Basin by 110 'polder pumping stations', with a total capacity of 48 m³/s. High-lying polders discharge via fixed weirs into the storage basin. Water from the storage basin is discharged via six 'main pumping stations', with a combined capacity of 54 m³/s, into the North Sea, the Nieuwe Waterweg and the Nieuwe Maas (Fig. 7.7).

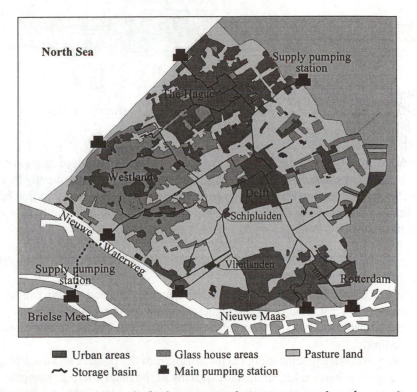

*Fig. 7.7. The Delfland Area (only the main regulating structures have been indicated).*

Glasshouse horticulture in the Delfland Area requires a large quantity of fresh water for irrigation purposes, especially in summer. Precipitation is collected from the roofs of the glasshouses and stored in rainwater basins (Fig. 7.8; see also Sec 4.2.1). This water is used for irrigation in the glasshouses. In summer, these rainwater basins may fall dry. In that case, surface water and drinking water are used for irrigation, sometimes in combination.

In agricultural areas such as pastures, where cattle farming is predominant, sprinkler irrigation is sometimes applied in summer. Surface-water use in general, requires the supply of water from the storage-basin to the polders. Therefore, water is let into the low polders via a large number of inlet structures. The majority of these inlets is small and is not frequently controlled. Water can be elevated up to the high-lying polders by polder pumping stations.

To maintain the target water level and the right water quality in the storage basin, a supply pumping station which has a capacity of 4 m³/s can be used. This pumping station abstracts water from the Brielse Meer (Fig. 7.7). Another large supply pumping station is located on the northern boundary of Delfland. It can let in water from the Rijnland Storage Basin and has a capacity of 8 m³/s. In the past, this pumping station was the only location where water could be let into Delfland. At present, the pumping station in the north is rarely used, because the quality of the water it supplies is less suitable than that of the Brielse Meer.

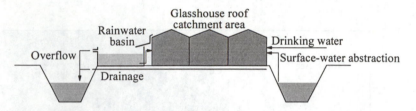

*Fig. 7.8. Rainwater basins in the glasshouse areas of Delfland.*

In summer, the quality of the surface water can become poor as a result of saline seepage and drainage water from glasshouses. The water quality is controlled by frequently flushing the entire water system. One of the purposes is to keep the chloride concentration below the official limit of 200 mg/l for irrigation in glasshouses. Depending on the type of crop, the practical limit may be lower.

The water level in the Delfland Storage Basin should be kept below NAP - 0.25 m. This level is called the 'milling-stop level'.

The resistance in the storage-basin canals, additional wind set-up and waves may result in water levels that are up to 0.3 m higher than average at specific locations. Since the height of the embankments along the storage basin is NAP + 0.1 m, this situation can almost lead to flooding.

During excessive precipitation, the discharge from the polders and the storage-basin land into the storage basin is higher than the capacity of the main drainage pumping stations. Therefore, the surface-water level may rise above milling-stop level at some locations in the storage basin. If the water level exceeds this limit, the Water Board can impose a milling stop at specific locations in the storage basin where problems are foreseen. In that case, certain polder pumping stations have to stop pumping, which may cause flooding of agricultural land. However, this occurs rarely.

During excessive precipitation, water can be stored in retention reservoirs temporarily. Two types of retention reservoirs are present in the area: reservoirs relieving the storage basin, called 'main retention reservoirs' and those relieving polders, called 'polder retention reservoirs' (Fig 7.9). Main retention reservoirs can only be used to store water from the storage basin, for instance, in the situation when the capacity of the storage basin turns out to be insufficient. When the water level in the storage basin falls, the retention reservoir can be emptied either by pumping or by gravity, into a low-lying polder. At present, the capacity of the main retention reservoirs is limited to approximately 0.02 m water height across the entire storage basin.

The polder retention reservoirs are used to store excess polder water, to prevent a milling stop, or during a milling stop, to prevent flooding.

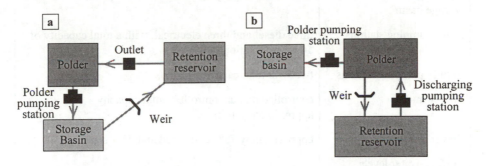

*Fig. 7.9. The principle of retention reservoirs in the Delfland Area: storage basin retention reservoirs (a) and polder retention reservoirs (b).*

In the Delfland Area it is common practice that urban areas drain to combined sewer systems primarily. The flows from these sewer systems are treated at sewage treatment plants (STPs), most of which discharge the effluent outside the water system. Via these routes, part of the precipitation is discharged from the water system, bypassing the storage basin.

During extreme precipitation, the combined sewer systems may overflow to the storage basin and polder surface water. This polluted water is discharged from these areas by polder pumping stations into the storage basin and from there by the main pumping stations to outside the system.

Important data on the Delfland water system, relevant for this case study, are listed in Table 7.2. For time-series calculations, hourly precipitation and evaporation data of Valkenburg and Rotterdam Airport meteorological stations of 1989 and 1990 have been used. Data on discharges of the main pumping stations per day and water-quality measurements once every two weeks have been made available by the Water Board. Further details on the Delfland water system can be found in Steenbekkers (1996).

Table 7.2. Delfland, data summary.

| Subject | Properties |
|---|---|
| Total area<br>main land use | 40,000 ha<br>41% rural, 28% urban, 19% glasshouse, 12% other |
| Types of areas | 72% low-lying polders, 6% high-lying polders, 22% storage-basin land |
| Number of polders | 57 |
| Main interests | flood prevention, ecology, glasshouse horticulture, pasture agriculture, recreation, navigation |
| Polder pumping stations pumping into to the storage basin | 110 draining low-lying polders; total capacity of 48 m³/s<br>11 draining high-lying polders; total capacity of 1.5 m³/s |
| Main pumping stations | three diesel and three electrical, with a total capacity of 54 m³/s (situation 1995) |
| Supply pumping stations | two with total capacity 12 m³/s |
| Polder inlets | controlled and uncontrolled, total capacity approximately 10 m³/s |
| Sewer systems | approximately 75% combined and 25% separate |
| Surface-water levels<br>  target storage basin<br>  milling stop<br>  target polders | <br>NAP - 0.40 m<br>NAP - 0.25 m<br>varies, for comparison purposes: NAP - 2.00 m |
| Capacity of retention reservoirs | 170,000 m³ main retention reservoirs; 1,400,000 m³ polder retention reservoirs |
| Chloride concentration limit for horticulture | 200 mg/l |

## 7.2.3  Water Management

### Current Practice

All polder-pumping stations in the Delfland Area are automated and operate on the basis of water-level setpoints in the polders. There is no central facility as yet to control all these pumping stations. The main pumping stations are all manually controlled. If the water level in the storage basin becomes too high, these pumping stations have to be manned. Sometimes operation of these pumping stations is also required during the night and in the weekend. It is common practice for the main pumping stations, to pump out water from the storage basin in advance. This creates an excess storage capacity and reduces the chance of having to operate outside office hours.

As shown in Fig. 7.10, the target water level of NAP - 0.40 m in the storage basin, is not maintained accurately at present. This is the result of the described practice to pump out water in advance. The graph shows that the largest water-level falls occur during and after periods of precipitation. This is partially the result of hydro-dynamic effects in the storage basin that occur during drainage, but also a result of overreacting to the fast-rising water levels in the storage basin during and after precipitation.

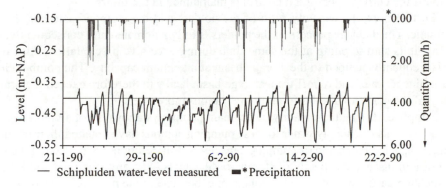

*Fig. 7.10. Water level measured in the Delfland Storage Basin at Schipluiden.*

To maintain the right water quality for irrigation purposes, flushing the water system is practiced frequently in summer. Many of the inlet structures to polders are manually controlled or kept open during the entire summer season. Thus flushing is continuous, possibly to a lower extent, but even during precipitation. It is the general opinion that as a result of this practice, the total quantity flushed is too large. The storage basin is flushed entirely, refreshing the surface water. This water is subsequently let in to the polders to flush them. The flushing rate for polders is generally larger than that for the storage basin.

The Water Board is currently reconsidering the design of polder systems. An important reason is the increasing urbanization, for which separate sewer systems are used, resulting in a faster runoff. The current policy is that surface-water storage and polder pumping capacity will have to be updated. Polder pumping stations that have too low a capacity, may require the building of larger pumping units, extension of the pumping station or an entirely new pumping station. Another option considered, is widening canals to create a larger storage capacity in the surface-water subsystem. All these measures are very costly.

Furthermore, the capacity of the storage basin and its main pumping stations are reconsidered. At present, main retention reservoirs are under construction, which will temporarily store excess water from the storage basin. Polder retention reservoirs do exist at present, but are rarely used.

Some of the main pumping stations already have too high a capacity to gradually lower the water level in the storage-basin canals. This results in a sudden fall in water level in the canal sections that supply the main pumping stations. If the main pumping stations would be controlled locally, such a fall can cause intermittent switching of pumping units.

### Problem Formulation

The increasing urbanization in the Delfland Area results in expansion of the impermeable area. Therefore the rapid discharge characteristics of the area are increasing. A rapid water-level rise in the polders and the storage basin, will lead to a lack of pumping or storage capacity, if the current operational control is maintained in the future.

Furthermore, the capacity ratio of the polders and the storage basin may become problematic. The drainage process of the polders is fully automated and consequently, most pumping units start to pump at the same time during excessive precipitation. As a result, water is frequently pumped to the storage basin at maximum capacity. The combination of polder drainage water and runoff from storage-basin land will therefore exceed the draining capacity of the main storage-basin pumping stations more often.

Simply extending the capacity of the main pumping stations is not a viable solution, since some canals that supply water to these pumping stations have reached their maximum discharge capacity. Extension of pumping capacity will therefore have to be combined with an increase in discharge capacity of storage-basin canals. At many locations in the storage basin such a measure would be very problematic, because of the present infrastructure and because the water levels of these canals are often far above the surrounding polders.

At present, rainwater basins in glasshouse areas store the initial precipitation. After these reservoirs have been filled, they overflow to the surface water. The moment the reservoirs are filled, the total area running off to the surface water, increases dramatically.

At present, glasshouse farmers have to install a storage capacity of at least 500 m³/ha. However, one of the future options for glasshouse farming is to produce irrigation water of good quality by means of reverse osmosis, using drinking water only. This enables closed systems that recirculate water and do not necessarily require rainwater suppletion. Such a development could make the current rainwater basins obsolete. If the rainwater basins are removed from the glasshouse areas, the total runoff from impermeable areas will increase permanently.

The present intensive use of surface water for irrigation purposes, forces the Water Board to let in a large quantity of alien water in dry periods. The concentrations of pollutants in surface waters are increasing continuously, as a result of recycling of surface water that is drained from glasshouses and pastures. Especially in dry summers, this process causes the water-quality limits to be exceeded. Therefore, the entire water system is flushed regularly. Since a large number of inlets is uncontrolled and stays open throughout the year, the total inlet of alien water supplied via the storage basin is large. However, the policy of the water board pledges to restrict the inlet of alien water to a minimum.

*Aim of the Case Study*

The case study aims at determining the options for enhanced water-system control in the Delfland water system, in the present situation and in the future. Of special interest is the method for reducing the number of problematic events resulting from extreme hydrological conditions, which should reduce the need for costly capacity enlargements.

Furthermore, the study aims at determining the quantity of inlet water, required to maintain the desired water quality. Special points of interest are increasing chloride concentrations in surface water during summer and surface water contamination resulting from combined sewer overflows.

The idea of the study is that, by balancing the sometimes conflicting interests in the water system, the existing resources can be used better to meet the various control objectives. In the present case study, the following objectives for control are set:

- to prevent flooding by water from the storage basin;
- to preserve the ecological values in the surface-water system as much as possible;
- to maintain required conditions for horticulture farming;
- to maintain required conditions for pasture agriculture;
- to maintain required conditions for recreation;
- to minimize operational costs.

## 7.2.4  Water-System Analysis

*Parameter Determination*

Detailed modeling of the Delfland water system requires detailed data on the many polders and regulating structures, however, these are not available at the moment. Therefore, the choice has been made to obtain the runoff characteristics of the polders in the Delfland water system by means of detailed analyses of two typical polders: a mainly rural polder (the Duifpolder) and a mainly glasshouse polder (the Woudse Droogmakerij). The following descriptions should be seen as examples of the results of the calibration and verification procedures described in the introduction of this chapter (Sec. 7.1.4).

For both polders, monitoring data have been gathered to determine their behavior. Parameters that influence the runoff characteristics of rural areas have been determined using a model of the Duifpolder. Parameters that influence the water quality behavior in glasshouse areas have been obtained from a model of the Woudse Droogmakerij.

The Duifpolder is an almost entirely rural polder, in which land-use, drainage capacity and soil texture are characteristic for most pasture areas in Delfland. A summary of data for this polder is listed in Table 7.3. Measuring took place in October and November 1994. Precipitation and evaporation data were recorded at Naaldwijk meteorological station, which is close to the Duifpolder.

Table 7.3. Duifpolder, data summary.

| Subject | Properties |
|---|---|
| Total area | 370 ha |
| Main land use | 94% rural, 1% urban, 1% glasshouse, 4% other |
| Soil-surface level | NAP - 2.77 m |
| Pumping-station capacity | approximately 0.63 m³/s |
| Time-series data | hourly precipitation and evaporation data from Naaldwijk monitoring station; hourly surface-water-level measurements at a representative location in the polder; manually registered pumping hours |

Figure 7.11 shows various variables, measured and calculated. The figure clearly shows that the fluctuations in the surface-water level modeled, correspond well with the water-level measured. Of special interest are the slopes in surface-water levels in graph (a). After precipitation, the rising water levels determine the runoff characteristics of the area, while

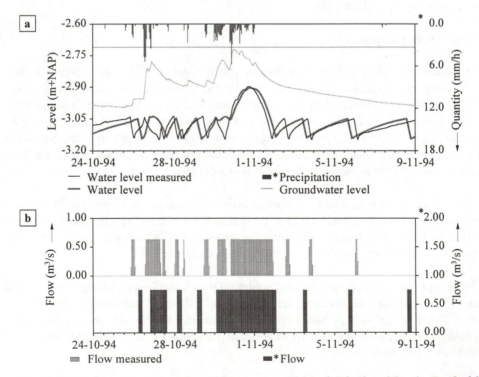

Fig. 7.11. Results of water levels and flows, measured and calculated for the Duifpolder.

falling water levels indicate the actual capacity of the pumping station. These were found to match.

Furthermore, graph (b) clearly indicates that the total quantity of water pumped from the system and the moments of pumping, correspond very well. However, when a water balance is drawn up, it can be proven that the flows measured at the pumping station are slightly too low. The difference between calculations and measurements could be due to several errors, as discussed in Sec. 7.1.4.

From the calibrated model, runoff parameters, such as soil resistance (Sec. 4.2.2), have been determined for the Duifpolder. It turned out that shallow gullies in the rural pasture areas give rise to very rapid runoff, once the groundwater level reaches the surface (not shown in the graph).

The Woudse Droogmakerij is a typical glasshouse polder. This specific polder has been chosen for water-quality simulation, because it almost entirely lacks saline seepage. Such seepage would affect the changes in water-quality, resulting from leaching from the glasshouse subsurface. The glass surface and the sizes of the rainwater basins have been determined on the basis of detailed maps of the area. Table 7.4 lists important modeling details of the Woudse Droogmakerij.

Table 7.4. Woudse Droogmakerij, data summary.

| Subject | Properties |
|---|---|
| Total area | 90 ha |
| Main land use | 28% rural, 10% urban, 55% glasshouse, 7% other |
| Pumping-station capacity | 0.27 m³/s |
| Inlet capacity | approximately 0.09 m³/s |
| Time-series data | averaged hourly precipitation and evaporation, based on data of Valkenburg and Rotterdam Airport monitoring stations; weekly chloride measurements in storage basin and the polder |

Graph (a) in Fig. 7.12 presents chloride concentrations measured and calculated for the Woudse Droogmakerij, for a period with low precipitation and high evaporation. Graph (b) shows water storage in rainwater basins.

During the period calculated, the chloride concentrations in the storage basin were used as boundary data. The inlets were kept open during the entire period, to enable dilution of polder water with water from the storage basin.

The general pattern of chloride development calculated for the Woudse Droogmakerij, closely resembles the actual measurements. The measuring data and the model calculations show that the water-quality limit of 200 mg/l for chloride is exceeded.

The example of the Woudse Droogmakerij demonstrates that during periods of drought, the water supply from the rainwater basins runs out. From that moment on, farmers start to use large quantities of surface water. Water without chloride evaporates, while the chloride leaches back to the surface water. This causes the chloride concentration in the polder to rise.

While interpreting the results, it should be kept in mind that the calibration is based on very few chloride measurements. The measurements show chloride concentrations at specific moments at one location in the polder, while the model describes the average concentration in the entire surface-water subsystem of the polder. The two values should therefore be compared with care.

The graphs make it clear that precipitation has quite a large impact on the chloride concentration in the surface-water subsystem. After each precipitation event, the chloride concentration clearly falls. The calculated curve does not exactly follow the measured chloride concentration. Of special interest are the concentrations measured on 2-8-90 and 8-8-90, which turn out to be the same. For that period, the model calculated a rise, followed by a fall in concentration. The reason for not measuring any change in the actual chloride concentration, is the precipitation that fell on 6-8-90, in between the moments of measuring.

Calibrating the model, parameters that determine changes in water quality in the surface-water subsystem were obtained, e.g.: the use of drinking water for irrigation, the quantity of water let in; the leaching fraction in glasshouse irrigation (Sec. 4.2.1 and 4.2.2).

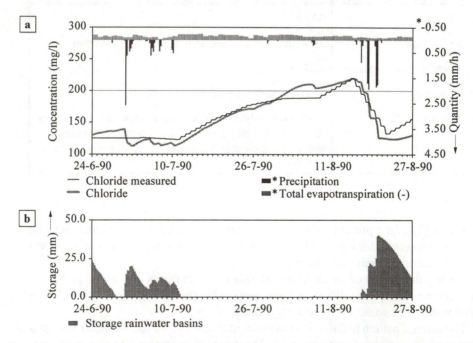

*Fig. 7.12. Results of chloride concentrations and related variables, measured and calculated for the Woudse Droogmakerij.*

*Water-System Modeling*

Two models of Delfland have been built to simulate the water system and its various modes of control (Fig. 7.13 and 7.14):
• a detailed model;
• a generalized model.

The detailed model is used to accurately describe the various subsystems and structures of the water system, including the geographical distribution of water-system elements. However, polders that have similar characteristics have been combined in the model.

The model has been verified using measured water-level extremes of 1990 and using data compiled from the operation of main pumping stations. Using this model the water-management practice of 1989 and 1990 has been calculated. Surface-water dynamics have been included in this model to examine the effects of water-level fluctuations as a result of operation of the main pumping stations. The retention reservoirs were not used in the verification period and therefore they have not been included in the detailed model.

The generalized model is used specifically to analyze the various combinations of control modes for the various structures regulating the water system. The reason for the generalization is to keep the optimization model to a workable size, while not violating the characteristics of the calculated runoff processes. The generalized model includes retention reservoirs for the storage basin and the low-lying polders. The geographical distribution of water-system elements is neglected; the storage-basin canals have been modeled together as one large reservoir.

In the generalized model, a distinction has been made between three types of low polders with typical runoff characteristics: rural polders, urban polders and glasshouse polders. The high-lying polders have been combined into one polder.

General data of the two models are listed in Table 7.5 (see Chapters 4 and 6 for details about the modeling of subsystems and structures).

Table 7.5. General data of the Delfland water-system models.

|  | Detailed model | Generalized model |
|---|---|---|
| Areas and Boundaries | 13 | 10 |
| Subsystems simulation | 37 | 25 |
| Subsystems optimization | - | 12 |
| Regulating structures | 21 | 15 |
| Fixed structures | 1 | 6 |
| Canal elements | 4 | 0 |

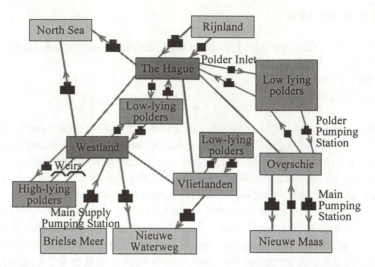

*Fig. 7.13. Schematic representation of the detailed Delfland model.*

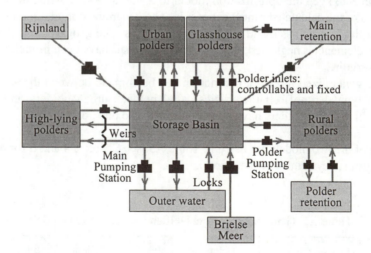

*Fig. 7.14. Schematic representation of the generalized Delfland model.*

### Simulating Current Practice

The years 1978 to 1993 have been analyzed in detail to determine representative periods to use in water-system simulation and verification (Lobbrecht, 1993b). The main considerations in selecting hydrological periods to use were: one period which exhibits average hydrological conditions; and another period which had a wet spring and a subsequent very dry summer followed by sudden and extreme precipitation events. Two consecutive years were found to fulfil these wishes: the year 1990, which matches the 60-year average hydrological conditions best and the year 1989, which showed the desired hydrological fluctuations and extremes. The Delfland models have been calibrated for the year 1989 and verified for the year 1990.

First of all, the monitored data were verified generally, followed by an overall calibration of the models, using water balances. By means of these water balances, overall inflows and outflows have been determined from the only available data, being the total quantity of water pumped from and into the storage basin (Steenbekkers, 1996).

The detailed and generalized models were further calibrated on the basis of net drainage data from the storage basin. This net drainage has been obtained by subtracting the flow of the main supply pumping station from the total discharge of the six main pumping stations that drain the storage basin. Figure 7.15 shows the actually measured net drainage from the storage basin and the net drainage determined by the detailed and generalized models.

Differences between measured and modeled drainage can be explained by the general effects discussed in Sec. 7.1.4: modeling inaccuracies and measuring inaccuracies.

Building a detailed model comprising every single polder and pumping station in Delfland, was considered outside the scope of the case study. This would have required a detailed survey, since most of the data required are not available. For this reason modeling inaccuracies are inevitable.

Local precipitation effects around monitoring stations may cause large differences in actual runoff and modeled runoff, especially in large areas in the summer season.

The draining capacity of main pumping stations is determined on the basis of running hours of engines, using average capacities. However, the actual capacity of a pumping station depends on the elevation height, which mainly fluctuates as a result of the tides in the North

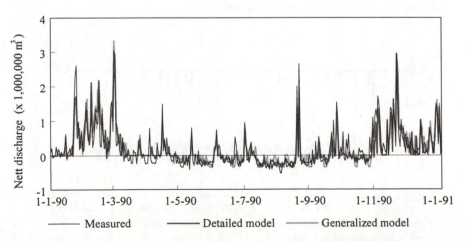

Fig. 7.15. Net discharge from the Storage Basin per day, based on manual control with quantities measured and local automatic control; with quantities calculated using the detailed and the generalized Delfland models.

Sea, the Nieuwe Waterweg and the Nieuwe Maas.

Furthermore, in Fig. 7.15, results of manual control are compared with results of local automatic control. Because of the fundamental differences between these modes of control, the results only serve general purposes, to get a rough impression of the effect of storage-basin drainage and inlet of alien water.

Because of the differences in control modes between the models and reality, the results shown in the figure could only be verified in general terms, checking the moments of peak discharges and the total water balance. Both proved to match very well, especially when considering the fact that only the conditions at the start of simulation were predetermined and no corrections were made during the one-year time-series calculation.

Figure 7.16 shows the water-level fluctuations in the storage basin as a result of local automatic control of the main pumping stations. The figure shows water levels calculated for the Westland region, at some distance from the main pumping stations and for Vlietlanden, which is closer to a main pumping station (see Fig. 7.7).

In comparison with the measured water levels resulting from manual control (see Fig. 7.10), the calculation results show a more stable water level in the storage basin. This is mainly due to the local setpoints of the main pumping stations, which cause the pumping engines to switch off as soon as the target level in the supply canal has been reached. As a result of hydro-dynamic effects, the water level fluctuates more, the closer to a main pumping station.

Another interesting feature which can be observed in Fig. 7.16, is that high water levels in the Westland region occur for longer periods than in Vlietlanden. First of all, the runoff from the large glasshouse horticulture area in the Westland, is very rapid. Furthermore, the discharge from the main pumping station near the Westland region is small, while the capacity of the storage basin-canals that discharge to other main pumping stations is limited. These effects explain why surface-water level in the Westland region rises fast and remains high for long periods. Consequently, the Westland shows the highest absolute water levels of the entire storage basin after precipitation.

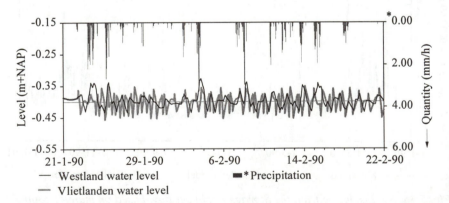

*Fig. 7.16. Water level in the Delfland Storage Basin for the Westland region and close to a main pumping station in Vlietlanden, calculated with the detailed Delfland model.*

*Interests Considered*

Several interests are considered for the Delfland water system: the common-good interests of flood prevention and ecology; the sectoral interests horticulture, pasture agriculture, recreation and navigation, and operational water-management interests.

The following key variables are distinguished in controlling the Delfland water system:
•     surface-water level,
•     chloride concentration,
•     BOD concentration.

Surface water levels play an important role in controlling the Delfland water system. As indicated above, especially in the areas along the storage-basin canals, flooding problems may occur when the surface-water level rises too high. To evaluate high water levels, the milling stop level of NAP - 0.25 m is used. Exceeding this level is here defined as 'storage-basin failure'.
      Two options can be distinguished which offer an improvement in control in general. The first option is to make better use of the retention reservoirs. Of utmost importance is the exact moment to start pumping into these reservoirs. If pumping excess water starts too late, too high water levels may occur, while the reservoirs are not even full. However, as mentioned before, the capacity of the main retention basins is rather limited. Furthermore, the polder reservoirs are hardly used at present, for various reasons. In this study it is assumed though, that the polder retention reservoirs are available and used.

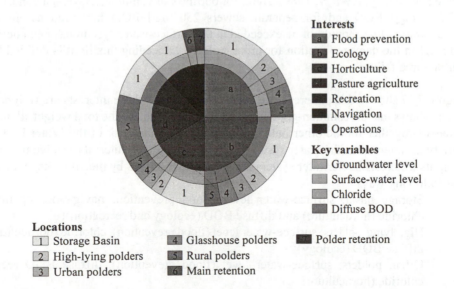

**Interests**
a  Flood prevention
b  Ecology
c  Horticulture
d  Pasture agriculture
   Recreation
   Navigation
   Operations

**Key variables**
   Groundwater level
   Surface-water level
   Chloride
   Diffuse BOD

**Locations**
1  Storage Basin        4  Glasshouse polders       7  Polder retention
2  High-lying polders   5  Rural polders
3  Urban polders        6  Main retention

*Fig. 7.17. Interest-weighing chart for the Delfland water system.*

The second option is to pump out water when extreme precipitation is expected. Water is pumped out of the polders into the storage basin and from there, out of the water system. This creates additional storage capacity in the storage basin.

In addition to flood prevention, the navigation interest depends on surface-water level control in the storage basin. This applies to both commercial and recreational navigation.

Too high chloride concentrations have disastrous effects on horticulture. In the current control practice, these disastrous effects are prevented by implementing high flushing factors for the polders. However, the Water Board aims at minimizing the inlet of alien water, which can only be accomplished if flushing is restricted to the minimum necessary. In dynamic control, it is assumed that the limit of 200 mg/l should not be exceeded. Therefore, this limit has been included in the surface-water damage functions of the storage basin, the glasshouse polders and the rural polders. Exceeding the limit is here defined as 'glasshouse-area failure'.

The current practice of frequently flushing the water system, has the advantage of removing undesirable pollutants such as BOD from the water system. This is important during and after combined sewer overflows, especially in summer. The BOD intended in this study is called 'diffuse BOD', to indicate its diffuse sources. The decay factor used for this type of BOD is 0.5 $d^{-1}$. The capacity of the inlets is so large that this pollutant can be flushed away before natural decay has reduced the pollutant load to an acceptable level.

If flushing the water system is only practiced on the basis of the chloride concentration, pollutants such as BOD will remain longer in the water system. To prevent this undesirable situation, the diffuse BOD-variable is included in determining dynamic control strategies.

Predicting the water-quality development in surface waters as a result of combined sewer overflow and separate sewer outflow is very difficult. This problem has not yet been solved accurately (NWRW, 1989). Here, for outflows of combined sewers, a concentration of 500 mg/l BOD and for separate sewers 250 mg/l BOD has been assumed. The concentration that should not be exceeded in the surface-water system has been defined as 20 mg/l in the damage function for urban polders. Exceeding this limit is defined here as 'urban-area failure'.

Figure 7.17 shows which interests are considered and how these interests are weighed. The figure shows that the common-good interests account for half of the total weight, all the other interests together for the other half. As in reality, the main task of the Water Board is to satisfy all interests in the water system, whereas operational interests receive the smallest weights. Interests in the water system are related to locations by means of key variables in the following ways:

- Storage basin: surface-water level (flood prevention, navigation, operational), chloride (horticulture) and diffuse BOD (ecology and recreation);
- High lying-polders: surface-water level (flood prevention), chloride (horticulture) and diffuse BOD (ecology);
- Urban polders: surface-water level (flood prevention), diffuse BOD (ecology), chloride (horticulture);

- Glasshouse polders: surface-water level (flood prevention), chloride (horticulture), diffuse BOD (ecology);
- Rural polders: surface-water level (flood prevention), chloride (horticulture) and diffuse BOD (ecology);
- Retention reservoirs: surface-water level (operational).

Figure 7.18 shows some examples of the damage functions used in the generalized Delfland model. Function (a) represents the damage to the flood-prevention interest. Flooding is considered to occur when the water level reaches NAP + 0.1 m. The preferred level in the storage basin is equal to the current target water level, at NAP - 0.40 m.

Function (b) presents the damage to ecology, here related to diffuse BOD concentration and mainly resulting from combined sewer overflows. The limit of the permissible level of BOD is 20 mg/l. Above this value, damage to the interest occurs, increasing with the concentration.

Function (c) presents the damage to the horticulture interest. This interest is related to chloride concentration and is considered to be seriously damaged if the concentration is over 200 mg/l.

Function (d) is a typical example of a navigation damage function. Once the water rises above a certain level, this interest is damaged, because boats may no longer go under the bridges. If the water falls below the lower level of the range, there is a risk of ships running aground.

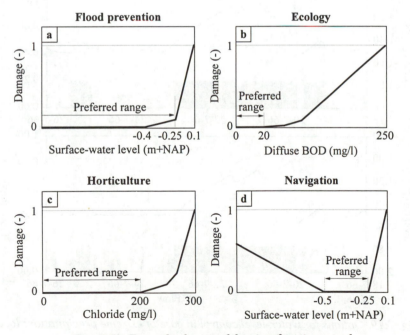

*Fig. 7.18. Example shapes of damage functions used.*

### Control Alternatives

Water-system performance in Delfland has been assessed by simulating the results of implementing various modes of control. The following modes are distinguished for controlling the regulating structures, in all possible permutations:
*   local manual control,
*   local automatic control,
*   dynamic manual control,
*   dynamic automatic control.

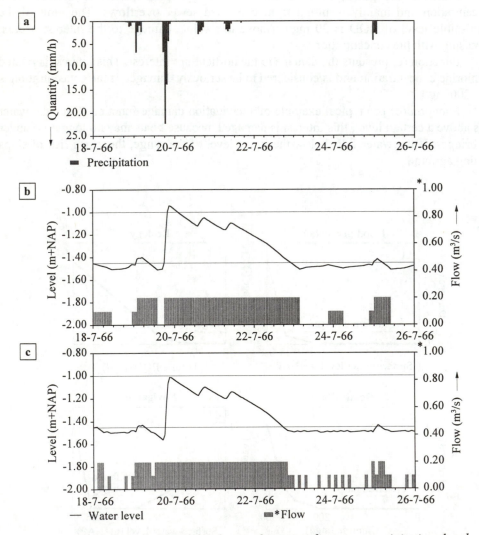

*Fig. 7.19. Options for water-level control in case of extreme precipitation: local automatic control (b: flooding), dynamic control (c: no flooding).*

Manual control of the main pumping stations, weirs and the majority of inlets, is the current practice. This type of control is used as a starting point to evaluate the various control options available.

Local automatic control is currently implemented in all polder pumping stations. Local automatic control will also be examined for polder-inlet structures.

Dynamic control is not applied at present. The effects of applying dynamic control, in the various types of regulating structures present in the Delfland Area is assessed here. Of special interest is the number of regulating structures that have to be dynamically controlled.

The effects of dynamic control in the Delfland water system are demonstrated below, starting with small-scale subsystem models and only one interest of which the requirements have to be satisfied. Subsequently, when the small-scale effects are explained, the effects of dynamic control on the entire Delfland water system are shown and analyzed.

*Flood-Prevention Interest*

At present, the Water Board uses a 'reference polder' to assess the effects of alternative measures which could be taken to enhance the water-system performance. The main land use in this polder is glasshouse horticulture. A model of the reference polder has been built, called the 'reference model', using calibration results of the Duifpolder and the Woudse Droogmakerij. The reference model is used to demonstrate how extreme water-system loads can be anticipated, preventing damage to the interests present in the subsystems. Table 7.6 gives an overview of the general data applying to the reference polder.

In the mainly agricultural polders, one of the most important issues in water management is to keep the surface-water level below the tops of the embankments to prevent flooding. Flooding of these polder areas is defined as 'system failure' by the Water Board.

A failure frequency of once every ten years is permissible in the design. For this reason, in principle, the system may, on average, fail once every ten years. To demonstrate

Table 7.6. Reference polder, data summary.

| Subject | Properties |
|---------|------------|
| Total area | 100 ha |
| Main land use | 30% rural, 5% urban, 60% glasshouse, 5% other |
| Target water level Soil surface level | NAP - 1.45 m NAP - 1.00 m |
| Pumping-station capacity | 0.2 m$^3$/s |
| Inlet capacity | 0.08 m$^3$/s |
| Time-series data | De Bilt, one hour resolution |

how the process of failure works and how it can be prevented by means of dynamic control, the extreme precipitation of 19 July 1966 serves as an example (Table 5.1, Page 95). That day, a total of 48.1 mm precipitation fell in only five hours (graph (a) in Fig 7.19, Page 186).

Graph (b) in Fig. 7.19 shows the calculated water level and drainage results, using local automatic control. As can be expected, very soon after that precipitation event, the surface-water level starts to rise. Within a few hours, the surface water level has risen that high, that the soil surface floods and the system fails according to the Delfland standards.

Graph (c) in Fig. 7.19 shows the surface water-level and drainage results which could be obtained using dynamic control. In this particular example, dynamic control takes action to prevent system failure, pumping approximately 0.12 m water from the surface-water subsystem in advance. This turns out to be just enough to prevent flooding.

Dynamic control anticipates system failure as a consequence of precipitation loads on the water system during the control horizon. However, in general, the weather forecast available is not sufficiently detailed to predict such an extreme, and probably only local

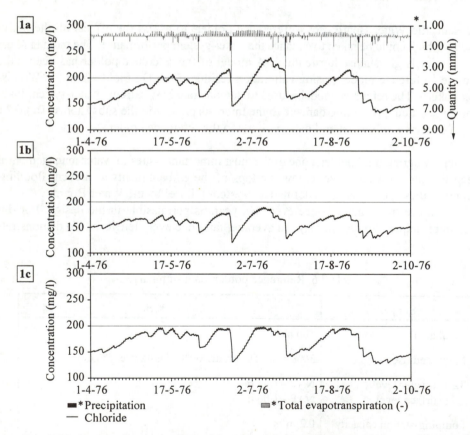

Fig. 7.20. Options for controlling chloride concentration in glasshouse polders: inlets permanently closed (1a), inlets permanently open (1b), dynamic control / inlets opened when required (1c).

precipitation event as the one of 19 July 1966. To include inaccuracies in prediction, an underestimate has been used in the example, by an uncertainty multiplier of 0.4, in fact, simulating a forecast of 20 mm, falling in 10 hours (instead of the actual 48.1 mm in five hours).

*Horticulture Interest*

To demonstrate the results of dynamic water-quality control, the model of the Woudse Droogmakerij, described above, has been used. The purpose here, is to demonstrate that the amount of water let in can be reduced and still the chloride concentration can be kept below the limit of 200 mg/l, using dynamic control. The summer season of 1976 has been selected for the analysis. This year was exceptionally dry (see Fig. 5.1, Page 92) and the summer season had a continuous, extremely dry period (Table 5.2, Page 97). The upper limit for the

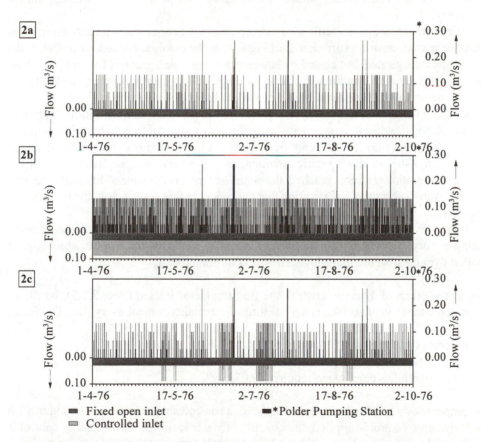

*Fig. 7.21. Options for controlling chloride concentration in glasshouse polders: inlets permanently closed (2a), inlets permanently open (2b), dynamic control / inlets opened when required (2c).*

chloride concentration in the Delfland Storage Basin has been kept fixed at 150 mg/l, which is the average value in a dry summer.

The entire inlet capacity (see Table 7.4), has been split into 0.03 m³/s for inlets that are kept open continuously throughout summer, irrespective of the conditions in the system, and 0.06 m³/s for controllable inlets. In combination with the permanently open inlets, for the controllable inlets the following options can be distinguished in the summer season:
a.      controllable inlets closed;
b.      controllable inlets open;
c.      controllable inlets open when required.

The results of the time-series calculations are presented in the graphs of Fig. 7.20 and 7.21. As shown in graph (1a) of Fig. 7.20, the chloride concentration regularly exceeds the limit of 200 mg/l, if the controllable inlets are kept closed. Graph (2a) of Fig. 7.21 shows that, despite the drought, the pumping station frequently discharges to the storage basin, which, combined with the open inlets, means the surface-water subsystem is flushed almost continuously.
        The current practice in Delfland is to keep most inlets open permanently throughout the summer season, ensuring sufficient flushing to keep the chloride concentration below the limit. Slightly exaggerated, but clear for demonstration purposes, graphs (1b) and (2b) show the result of this practice, continuously using the entire inlet capacity. As presented in graph (1b), chloride concentrations remain well below the limit.
        Graph (1a) shows that on five occasions, the chloride concentration limit was exceeded. During these periods apparently, extra flushing of the surface-water subsystem was required to keep the chloride concentration below the limit. This is exactly what dynamic control accomplishes in this specific situation, as can be seen in graph (2c). When the chloride concentration is almost reached, the controlled inlet is opened, additionally flushing the surface-water subsystem and ensuring that the concentration stays below the limit during the drought.
        In the dynamic control option described, a control horizon of 12 hours is used, assuming a 'dry' hydrological-load scenario (Sec. 5.3). Only the controllable inlet is controlled dynamically, the pumping station is controlled locally.

To assess the effect of dynamic control, the flushing factor is used (Sec. 7.1.5), which, in short, presents the ratio of surface water substituted for quality control every day. The factor turns out to be:
a.      controlled inlets closed: 0.04 day⁻¹;
b.      controlled inlets open: 0.16 day⁻¹;
c.      controlled inlets open when required: 0.05 day⁻¹.

This example shows that dynamic control achieves a considerable reduction in letting in alien water. If dynamic control is applied, the quantity of water let in for water-quality control is only 31% of the quantity that would have been let in, in the current practice.

*Comparing Control Alternatives*

Now the effect of dynamic control has been shown in two small-scale examples, the entire water system of Delfland will be discussed in the remainder of the following sections.

To compare local and dynamic control, a period of 30 years has been calculated, using the De Bilt time series for precipitation and evaporation, in time steps of 1 hour. In the local automatic control mode, all pumping stations are controlled locally. Large and controllable inlets are kept permanently closed in the winter season and permanently open in the summer season. Small inlets are open throughout the year. Weirs, draining the high-lying polders, are of the fixed type.

In dynamic control mode, all structures that are controlled locally in local automatic control mode are assumed to be controlled dynamically. Furthermore, large inlets are controlled dynamically. This situation will be called the 'basic alternative' below. Minimum operational periods are set to 6 hours for half the storage basin drainage capacity, to represent manually applied dynamic control.

A control horizon of 24 hours has been used, applying time steps of two hours and the following prediction scheme:

- 0 - 4 hours: period-average prediction, uncertainty multiplier 0.6;
- 4 - 8 hours: period-average prediction, uncertainty multiplier 0.6;
- 8 - 24 hours: period-average prediction, uncertainty multiplier 0.6.

To speed up the calculation process, a dynamic-control threshold has been used (Sec. 7.1.2). Using this technique, it turns out that dynamic control has to come into action only 6.6% of the time.

Table 7.7. Time-series calculation results of local automatic and dynamic control of the Delfland water system.

| | Performance index (-) | Storage Basin failure events $(y^{-1})$ / total duration (h/y) | Polder flushing factor $(d^{-1})$ | Glasshouse area failure duration (d/y) | Urban area failure duration (d/y) |
|---|---|---|---|---|---|
| Local Automatic Control | 1 | 0.13 / 1.27 | 0.043 | 7 | 10 |
| Dynamic Control (basic alternative) | 1.69 | 0.03 / 0.07 | 0.021 | 8 | 10 |

The most important results obtained assuming local automatic and dynamic control are given in Table 7.7. Because of the great variety of interests included in the model, the performance index has been used for comparison (Sec. 7.1.5).

First of all, Table 7.7 indicates that the current number of storage basin failures is 0.13, which is slightly above the criterion of 0.1 set by the Water Board. Furthermore, the table shows that performance improvement using dynamic control instead of local automatic control is 69%.

The number and duration of storage-basin failures can be reduced using dynamic control, which means that the surface-water level in the storage basin is better maintained below the milling stop level during extreme precipitation events. This result has been accomplished with a reduction in main retention reservoir use from 0.83 a year using local automatic control to 0.60 a year using dynamic control (not shown in the table).

The calculations further show that the total quantity of water used for flushing the polders, can be reduced by more than 50%. This result is obtained, with admittedly a slight increase in the duration of glasshouse failures from seven to eight days a year. The reason for this phenomenon is the fact that in dynamic control continuous flushing in summer is not practiced, as it is in local automatic control. Dynamic control comes into action when the chloride concentration reaches the limit, but under extreme conditions, this is apparently too late. However, water-quality control in the urban areas does not show an increase in the duration of urban-area failures, the duration remains 10 days a year.

When comparing the results presented in Table 7.7, it should be kept in mind that only those failure events and durations are listed that represent extreme-system-load situations. The performance index incorporates all conditions of the water system during each time step calculated, including the majority of periods with low system loads. This means that, while the extremes of two possible alternatives may be comparable, the overall performance may differ. A higher performance indicates that the average conditions are satisfied better.

In the basic alternative, the majority of structures is dynamically controlled. The question arises, whether the results obtained using the basic alternative would be as good, if part of the water system is controlled locally. To determine the effect of controlling specific groups

Table 7.8. Alternatives to dynamic control of all the regulating structures in Delfland and control modes of various types of regulating structures.

| Regulating structures [1] | Alternatives [2] | | |
|---|---|---|---|
| | Basic alternative | Reduced alternative | Water-quantity alternative |
| Polder pumping stations | DC | LAC / DC | DC |
| Main pumping stations | DC | LAC / DC | DC |
| Main supply pumping station | DC | DC | LAC |
| Large polder inlets | DC | DC | LAC |
| Retention reservoir structures | DC | LAC / DC | DC |

[1] Small polder inlets and weirs are of the fixed type in all alternatives.
[2] LAC = Local Automatic Control mode; DC = Dynamic Control mode

of structures locally, the two alternatives presented in Table 7.8, have been examined.

Alternative (a), called the 'reduced alternative', describes a situation in which the number of dynamically controlled structures is reduced. Half the main pumping station capacity is controlled locally, the other half dynamically, while the polder pumping stations supplying the high-lying polders are controlled locally. In this alternative it has been tried to ensure good system performance and handling of extreme events.

Alternative (b), called the 'water quantity alternative', only focuses on water-quantity control. The purpose of this alternative is to determine whether the improvement in water-level control in the storage basin and the polders changes when water-quality control is neglected. It is similar to the basic alternative, but the main supply pumping station and large polder inlets are locally controlled.

The results of dynamic control according to these two the alternatives are given in Table 7.9.

The reduced alternative (a) shows a less favorable performance index than the basic alternative. This is due to the fact that half the main pumping stations' capacity is controlled locally. Furthermore, the number of storage-basin failures does not increase, while the duration of these failures does increase. The durations of urban-area and glasshouse-area failure events are similar to that of the basic alternative.

The water-quantity alternative (b) shows only a very small reduction in performance index in comparison to the basic alternative. Since no conflicting interests are incorporated in control, dynamic control results in no storage basin failures during the 30-year period calculated. The flushing factor, however is as poor as when using local automatic control.

Table 7.9. Time-series calculation results of dynamic-control alternatives of the Delfland water system.

| | Performance index (-) | Storage Basin failure events $(y^{-1})$ / total duration (h/y) | Polder flushing factor $(d^{-1})$ | Glasshouse area failure duration (d/y) | Urban area failure duration (d/y) |
|---|---|---|---|---|---|
| Dynamic Control (basic alternative) | 1.69 | 0.03 / 0.07 | 0.021 | 8 | 10 |
| Dynamic / Local (a; reduced) | 1.48 | 0.03 / 0.27 | 0.022 | 8 | 10 |
| Dynamic / Local (b; water-quantity) | 1.67 | 0.00 / 0.00 | 0.043 | 6 | 10 |

### Future Developments in the Area

To determine the effect of possible future developments in the Delfland Area, local automatic control and dynamic control according to the basic alternative, have been examined more closely, using two different development scenarios.

In the first scenario (A), the effect of urban expansion has been analyzed, assuming an increase of approximately 15% of the entire urban surface, which is the predicted expansion over the next 10 years. All new urban areas are assumed to be drained by separate sewer systems.

In the second scenario (B), in addition to urban expansion, all rainwater basins are assumed to have been removed from the glasshouse areas. This implies a reduction in extra storage, which in the current situation is very effective in summer, when farmers require large quantities of water for irrigation. Under the current government policy this is a hypothetical situation, but it is included to demonstrate the effect of possible future changes in policy.

Table 7.10. Time-series calculation results of local automatic and dynamic control for two development scenarios, incorporating possible future expansions in Delfland.

| | Performance index (-) | Storage Basin failure events $(y^{-1})$ / total duration (h/y) | Polder flushing factor $(d^{-1})$ | Glasshouse area failure duration (d/y) | Urban area failure duration (d/y) |
|---|---|---|---|---|---|
| Dynamic Control (basic alternative) | 1.69 | 0.03 / 0.07 | 0.021 | 8 | 10 |
| Local Automatic Control (Scenario A) | 0.87 | 0.30 / 1.33 | 0.043 | 5 | 14 |
| Dynamic Control (Scenario A) | 1.38 | 0.03 / 0.07 | 0.021 | 6 | 15 |
| Local Automatic Control (Scenario B) | 0.73 | 0.30 / 1.80 | 0.043 | 4 | 14 |
| Dynamic Control (Scenario B) | 1.27 | 0.03 / 0.13 | 0.021 | 4 | 15 |

Table 7.10 shows that applying local automatic control will in future reduce the performance index to 0.87 if only urban expansion (A) is taken into account, and to 0.73 if, in addition, the rainwater basins are removed (B). Furthermore, the number of storage basin failures increases to 0.3 events a year, which is three times as high as the permitted number.

The performance index can be improved in both development scenarios, by applying dynamic control. Dynamic control would also reduce the number of storage-basin failures to 0.03 events a year.

The duration of glasshouse-area failures in the future scenarios is shorter than in the current basic alternative (Table 7.7). This is due to the increase in runoff from urban areas after precipitation events, which contributes to flushing the surface water.

The number of urban-area failures increases in all future scenarios, as can be expected. The total quantity of pollutants discharged to the surface water, will increase as a result of the separate sewer systems.

### Extremes Calculation

To find out what results can be obtained by applying dynamic control for the entire water system, this subsection presents detailed time-series calculation results. The results of local and dynamic control are presented for an extreme situation: a heavy rainstorm, causing extremely high water levels. The winter rainstorm of 26 November 1983 (see Table 5.1, Page 95) has been selected for the purpose.

An important reason for selecting this particular event is the problem which the Water Board faced the next day: 27 November 1983. During and shortly after the event, the Water Board had to impose a milling stop for several polders and to switch off some polder pumping stations, to prevent storage-basin failure.

The results of water-level calculations during the event have been verified on a general scale. For the time-series calculations presented here, precipitation and evaporation data measured at Valkenburg and Rotterdam Airport meteorological stations have been used. The total precipitation was 49.0 mm, which fell in 21 hours, almost as much as recorded in De Bilt during that same event (Table 5.1). The preceding precipitation event had caused the soil to be saturated and the rainwater basins to be filled.

In 1983, the main retention reservoirs were not yet present in the Delfland water system. However, for demonstration purposes, the main retention reservoir is used in the example. The polder retention reservoirs are, on purpose, not used.

### Local Automatic Control

To prevent high water levels in the storage basin, a milling-stop level is set for the rural polder pumping stations in local automatic control. As can be observed in graph (b) of Fig. 7.22 (Page 196), during the precipitation event, the rural polder-pumping stations are switched off, as a result of a milling stop (6 hours on 27-11-83).

Furthermore, the main retention reservoirs (d) are used, trying to avoid the water level to rise above the milling-stop level. The water-level curves (c) show that the storage-basin failure lasts almost an entire day in the calculation. Furthermore, the water levels in the polders on average rise to 0.25 m above their targets.

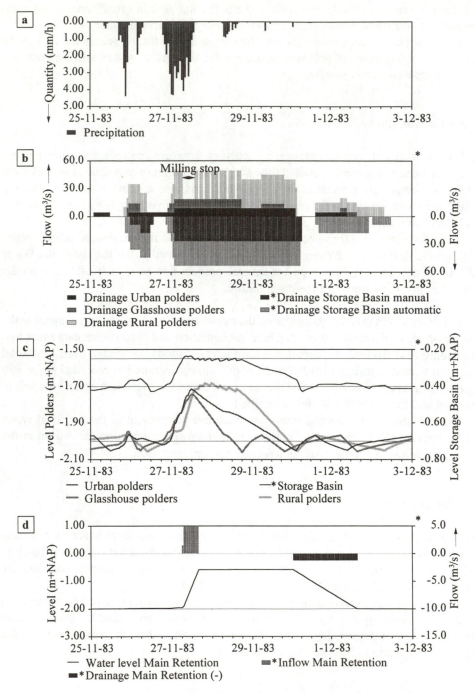

*Fig. 7.22. Time-series calculations for local automatic control during the rainstorm of 26 and 27 November 1983.*

*Dynamic Control*

In dynamic control, the prediction scheme for precipitation as presented before has been used, including an underestimate of the rainstorm. A 30 mm rain storm has been predicted, falling in 24 hours (instead of 49 mm in 21 hour).

When examining graph (b) of Fig. 7.23, it can be observed that all pumping stations start to operate slightly earlier than in local automatic control mode. Furthermore, it can be seen that dynamic control automatically applies a milling stop for the rural polder pumping

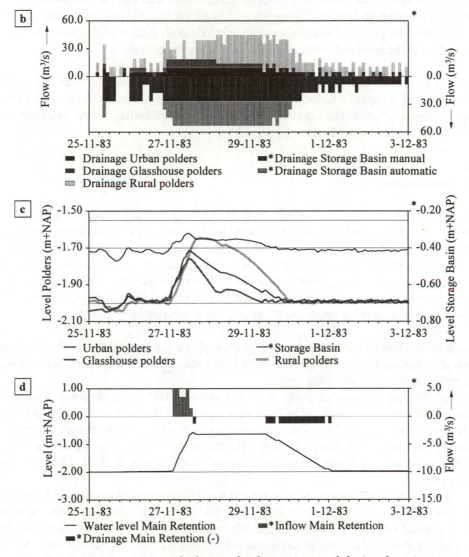

Fig. 7.23. Time-series calculations for dynamic control during the rainstorm of 26 and 27 November 1983.

stations, even when the failure level in the storage basin has not been reached yet. The reason for this strategy is to prevent too high water levels in the storage basin, while there is enough storage capacity left in the rural polders. This result could be expected, considering the weighing of the storage basin and the rural polders, presented in the chart of Fig. 7.17 (Page 183).

The precipitation prediction used, lacks the accuracy to start pumping well in advance. The water level in the storage basin is lowered just before the precipitation event, but only very little.

The total range of fluctuation of the storage basin water level is much less than that obtained by local automatic control. Two causes can be mentioned that resulted in this effect: starting the main pumping stations in time and keeping more water in the rural polders during the event. The rural polders therefore show an approximately 0.05 m higher water level on average, than in the local automatic control mode of the water system.

The overall picture which can be obtained from the time-series calculation example, is that dynamic control is superior to local automatic control because it makes better use of the available system resources. The extreme precipitation is handled better, without even needing an accurate precipitation prediction.


## 7.2.5  Conclusions

The increase in impermeable area, as a result of increased urbanization and expansion of the glasshouse area, forces the Delfland Water Board to reconsider the current system capacities and the way in which these capacities are used. In principle, all present and future problems related to a lack of system capacity, can be solved by increasing the surface area of the water bodies, by enlarging the storage-basin canals, by building new pumping stations and similar measures. However, such measures are very costly. The present case study presents a much cheaper alternative: making better use of all existing water-system resources.

The case study shows clearly that the storage capacity currently available in the Delfland water system has reached its limit, but it can be used better when dynamic control is applied in the entire or in part of the water system.

The 30-year time-series calculations demonstrate that using dynamic control, a significant reduction in the number of extremely high water levels in the storage basin can be achieved, by as much as 75%. Dynamic control can thus be used to enhance the current water-system performance, but also to maintain the safety of the storage basin, especially when future expansions of the urban area or other developments in Delfland cause even faster runoff.

In the glasshouse polders, the number and size of rainwater basins play an important role in the frequency of high water levels and the number of system failures. If these reservoirs would be abolished in future, the Water Board will have to take action, in order to at least maintain the current frequency of system failure. One of the options is to use dynamic control.

Depending on the weight of the interests present in the various subsystems of the water system, dynamic control will satisfy the requirements of these interests. The high weighing of the storage basin, as incorporated in the examples of this case study, shows that during heavy rain storms, priority is given to flood prevention in the storage basin; actively using the retention reservoirs.

The weighing of water-quality-related interests present in urban, glasshouse and rural polders, implies that when using dynamic control, the surface-water subsystems are only flushed when necessary. This would reduce the total quantity of alien water inlet by 50% on average. Both in local and dynamic control modes the chloride concentration standard of 200 mg/l would be violated during extreme drought. This is due to a lack of fresh water to flush the water systems and a lack of inlet capacity.

In case of surface-water pollution, resulting from combined-sewer overflows and separate-sewer outflows, dynamic control would only flush the surface waters in urbanized areas after the overflows occur. In these particular cases, and for water-quality control in the Delfland Area in general, no prediction of the water-system load is required, since control actions would be taken as soon as the water-quality variables violate the standards. The results of dynamic control and local automatic control are the same in this case.

The reduction in letting in water to the polders during the summer results in a general reduction in polder-water pumping. This will contribute to lower running costs for the electrical polder-pumping stations.

Dynamic control can be implemented in the Delfland water system in several stages. As demonstrated, in the reduced alternative, not all regulating structures have to be automated to obtain an improvement in water-system performance and reduce the number of water-system failures considerably.

The current automation configuration in the area already enables central data gathering of water levels in the polders and the storage basin and compilation of operational data of pumping stations.

The main elements of the water system that should be included in dynamic control are: large polder pumping stations, large polder inlets and three main pumping stations, representing approximately half the entire pumping capacity. For the current, larger polder pumping stations the option should be created to switch from the regular local automatic control mode to remote control by a central control unit.

Three of the six main pumping stations, could be controlled locally as described. The other three could still be controlled manually like they are at present. In these manually controlled pumping stations, dynamic control could be applied, introduced as dynamic manual control in Sec. 2.4. In that case, the operators should interpret and follow the control strategy determined by the Strategy Resolver.

In the various examples given, the option of manual implementation of dynamic control strategies has already been incorporated by ensuring a minimum operational period for the manually controlled main pumping stations of 6 hours.

## 7.3   De Drie Ambachten Case Study

### 7.3.1 Introduction

The Drie Ambachten area is located in the south-western part of the Netherlands, along the lower reaches of the Westerschelde river (see Fig. 1.1). The area is bordered in the north by the Westerschelde, in the south by Belgium and in the east and west by areas with comparable water-management characteristics. The authority responsible for water management is the Water Board 'De Drie Ambachten'.

The area is included in the present study since it demonstrates several water-control problems which are typical for gravity-drained polders along the coast and in tidal river deltas. A specific point of interest is the cross-border discharge of excess water from Belgium.

### 7.3.2 Water-System Description

The area of De Drie Ambachten comprises three independently controlled districts, with a total surface of 24,000 ha (Fig. 7.24). Excess water from the area is discharged to the Westerschelde via two systems of creeks and canals. These systems are separated by the 'Kanaal van Gent-naar-Terneuzen', which is an important shipping route from Belgium to the Westerschelde. District 1 is situated west of the canal; Districts 2 and 3 east of the canal. Districts 1 and 3 receive cross-border water from Belgium. Excess water from upstream and from District 1 is discharged to the Westerschelde via the Braakman Sluice. The discharge of this sluice is entirely gravity driven.

District 3 discharges water to District 2 via a large weir at the village of Axel. Upstream and excess water from District 2 is discharged via the Otheensche Sluice, a sluice which is equipped with pumping units. These units are used during periods when the drainage of excess water by gravity is impossible, because the water level of the Westerschelde is too high.

Water management in the area around the Braakman Kreek (District 1) is the main subject of this case study. This area is defined as the 'Braakman Area'. Its surface area is approximately 5,000 ha. The Braakman Area does not include the Kanaal van Gent-naar-Terneuzen and the canal parallel to it, which both discharge to the Westerschelde by separate sluices. The Braakman Area slopes gently and comprises a system of several polders, with approximately 30 different water-level sections, which discharge excess water to the Braakman Kreek via a surface-water system of small canals, weirs and a pumping station. For each polder, target levels have been defined for the summer and winter seasons.

The soils of the Braakman Area consist of silty clay and silty fine grained sand. The lowest polders have a surface level of NAP + 0.1 m. In the eastern part of the area, the surface level is up to NAP + 2.2 m: a sandy area.

The Braakman Area has predominantly agricultural interests (arable farming and horticulture), while interests of nature, recreation (swimming, sailing and fishing) are present in and adjacent to the Braakman Kreek.

The water system that drains to the Braakman Kreek is located for the larger part in Belgium. In this Belgian area, rural drainage water is collected in the Leopold Kanaal, which can discharge via a sluice near Heist, or via the Isabella Pumping Station, located at the national frontier (Fig. 7.25). Depending on the discharge possibilities of the sluice near Heist, the Isabella Pumping Station drains an area of approximately 18,000 ha and discharges its water into the Braakman Kreek. On an annual basis, this represents 65% of the total discharge into the Braakman Kreek.

A minor part of District 1 discharges into the Leopold Kanaal. For simplicity, the entire area discharging into the Leopold Kanaal is called the 'Belgian Area', although part of it is actually in the Netherlands.

The target surface-water level in the Leopold Kanaal is NAP - 1.00 m. The target levels in the Braakman Kreek are NAP - 0.90 m in the summer season and NAP - 0.40 m in the winter season. The prevailing interest in the area discharging into the Leopold Kanaal is agriculture.

Agricultural and nature areas in the eastern part of the Braakman Area suffer from drought in summer. This problem is dealt with in two ways: irrigation using water from the Braakman Kreek and irrigation using alien water. The effect of this irrigation is relatively small, however. One of the problems is that, given the current water-system arrangement, it is rather difficult to supply water from the Braakman Kreek. Another problem is that the water quality of the alien water is not constant, sometimes even very poor. In the modeling described below, the irrigation works are not considered.

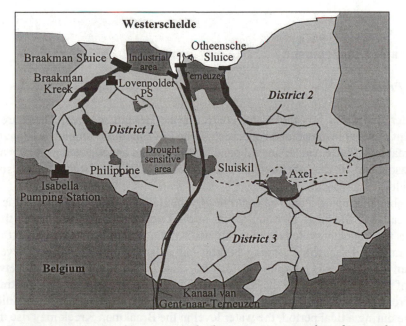

*Fig. 7.24. De Drie Ambachten area (only the main structures have been indicated).*

Details of the water system are listed in Table 7.11. Further details can be found in Janssens (1994).

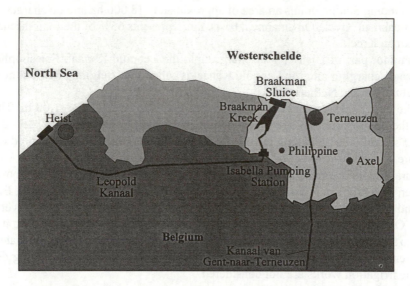

*Fig. 7.25. Discharge from the Leopold Kanaal: via the sluice near Heist and via the Braakman Sluice.*

### 7.3.3 Water Management

*Current Practice*

The Water Board has few means for control in the Braakman Area. These are: manual adjustment of weirs in the polders; operation of the Lovenpolder Pumping Station; operation of the Braakman Sluice. Under normal conditions the weirs are adjusted manually at the turn of the season. Only in case of extreme water-system loads they are occasionally also adjusted in the middle of a season.

Isabella Pumping Station is controlled by the Belgian water authorities. It is operated automatically on the basis of water levels in the Leopold Kanaal. However, the current practice is to frequently switch the pumping units of the pumping station on or off manually. The operation of Isabella Pumping Station has a large impact on water levels in the Braakman Kreek. If this pumping station is operated at maximum capacity between two periods of regular discharge by the Braakman Sluice, the water level in the Braakman Kreek rises approximately 0.2 m. As will become clear later, each time the Braakman Sluice cannot discharge during a tidal period, the water level in the Braakman Kreek may rise more.

Table 7.11. De Drie Ambachten, data summary.

| Subject | Properties [1] |
|---|---|
| Water-system area | Braakman Area: 5,000 ha, gently sloping; Belgian Area: 18,000 ha, flat; Braakman Kreek surface: 150 ha |
| Main land use | 75% rural, 5% urban, 4.5% surface water, 15.5% other |
| Soil-surface levels | NAP + 0.1 m to NAP + 2.2 m |
| Polder water levels | NAP - 1.4 m to NAP + 0.9 m |
| Tidal range Westerschelde | NAP - 2.04 m to NAP + 2.03 m |
| Hydrological seasons | summer: 1 March to 30 September, winter: 1 October to 28 February |
| Target surface-water levels | Braakman Kreek: NAP - 0.40 m in summer, NAP - 0.90 m in winter Leopold Kanaal: NAP - 1.00 m |
| Drainage system general | field canals: 1 m wide; discharge canals: 4 m wide, approximately 30 fixed weirs; Lovenpolder Pumping Station with capacity 2.9 m$^3$/s |
| Braakman sluice | 4 openings: 24 m$^2$ total area; miter gates and sliding gates; maximum discharge capacity approximately 60 m$^3$/s |
| Isabella Pumping Station | 6 pumping units, total capacity 13.2 m$^3$/s |
| Time-series data | water-level measurements of 1992, once a day; precipitation totals per day, measured at Philippine; monthly discharge totals of Isabella Pumping Station |

[1] Data from the Braakman Area, if not specified.

The fixed weirs in the Braakman Area are set in spring at the high level and in autumn at the low level. The purpose is to preserve as much water as possible during summer and have sufficient storage capacity in the canals during winter.

Lovenpolder Pumping Station is controlled automatically on the basis of its upstream surface-water level.

The Braakman Sluice is currently operated by remote manual control of the sliding gates. Miter gates close the sluice when the water level of the Westerschelde is above that of the Braakman Kreek (Fig. 7.26).

In general, in summer the evaporation exceeds the precipitation and therefore the Braakman Sluice may be closed for long periods to preserve water. In such periods, deep and shallow saline seepage, cause the chloride concentration in the Braakman Kreek to rise.

An automatic water-quantity and water-quality monitoring network is currently under construction. The purpose is to gather data about the water-system state to improve water management. One of the study topics is whether information from the Belgian Area can and should be incorporated in the monitoring system, especially data on operation of the Isabella Pumping Station.

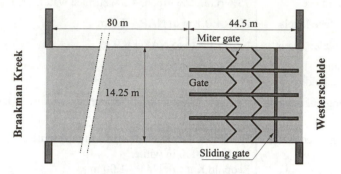

*Fig. 7.26. Top view of the Braakman Sluice.*

### *Problem Formulation*

The various interests present in and along the Braakman Kreek all require a stable surface water level. Because the slope of the shores along the Braakman Kreek is very gentle, at some locations, small variations in water level greatly affect the available recreation area. High water levels reduce the width of the beach strips. Yacht jetties may flood as a consequence of too high water levels.

The Isabella Pumping Station is entirely controlled by the Belgian authorities and there is no communication on its operation with the Water Board at all. If necessary, the pumping station is simply switched on. This can cause severe problems, especially when it is not possible to discharge via the Braakman Sluice. This situation occurs during continuous severe storms on the North Sea, during which the low-tide water level of the Westerschelde stays too high to enable gravity discharge. Since there is no other means to discharge excess water, the water level in the Braakman Kreek continues to rise until a sufficiently positive water-level difference occurs. Figure 7.27 shows an example of this problem around 20-11-1992.

Furthermore, manual control of the Braakman Sluice causes large fluctuations in water level. Sliding-gate openings are generally fixed for several hours or days, depending on the actual water-system runoff. Despite adjustment of the sliding gates, manual control does not enable accurate water-level control.

The Water Board aims at meeting the requirements of agricultural, nature and recreational interests as well as possible, by maintaining surface and groundwater levels at their target levels. At present, the number of automatically controllable structures in the area is that low, it is not possible to preserve water in the higher polders in dry summers.

As a result of stagnant water and saline seepage, water quality in the Braakman Kreek becomes poor during dry summers.

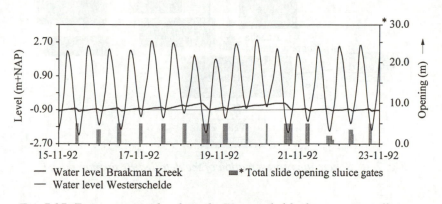

Fig. 7.27. Extreme water levels in the Westerschelde that restrict sufficient discharge via the Braakman Sluice.

*Aim of the Case Study*

In this case study the options for improved water-level control in the Braakman Kreek are investigated. The general idea is that the Braakman Sluice can be controlled, anticipating runoff from the water system. The objective formulated for control, is to keep the water level in the Braakman Kreek as stable as possible in both the summer and winter seasons.

In the summer season, recreational interests prevail. However, operational interests have to be satisfied in that season as well. In the winter season the flood-prevention interest prevails, while the operational interests are still present. The flood-prevention interest requires maintaining the winter target water level accurately, ensuring a safety margin for periods during which discharge via the Braakman Sluice is restricted or impossible as a consequence of high water levels in the Westerschelde. Water coming from the Isabella Pumping station should be stored in the Braakman Kreek in this situation until discharge via the sluice is possible again.

Discharging water via the Braakman Sluice into the Westerschelde is most often restricted in winter. Therefore, this period imposes the largest restrictions on water management in the area in general.

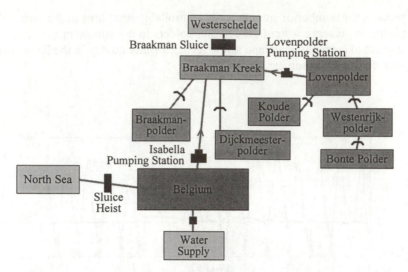

*Fig. 7.28. Schematic representation of the detailed Braakman model.*

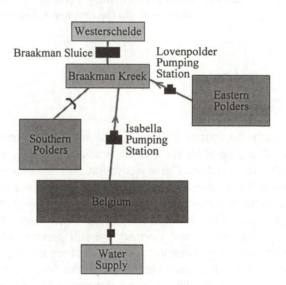

*Fig. 7.29. Schematic representation of the generalized Braakman model.*

### 7.3.4 Water-System Analysis

*Water-System Modeling*

The Braakman Kreek plays a central role in the discharge process of the water system. The Braakman Kreek acts as a reservoir that collects water from its environment. Maintaining the water level in the Braakman Kreek has a limited effect on groundwater and surface-water levels in other parts of the water system.

Two models of the Braakman water system have been built to simulate runoff processes and control (Fig. 7.28 and 7.29):
1.    a detailed model;
2.    a generalized model.

The first model describes the various subsystems and structures of the water system in detail. The model has been verified using water-level measurements of 1992 and information compiled about the operation of the Isabella Pumping Station. Using this model, various system parameters have been determined, to describe the current practice as well as possible.

The second model has been used to analyze the control options. The reason for generalization is to keep the optimization model to a workable size, while the characteristics of the calculated processes are not violated.

General data of the two models are listed in Table 7.12 (see Chapters 4 and 6 for details on modeling of subsystems and structures). In both models the main controllable structures are: Isabella Pumping Station, Lovenpolder Pumping Station and the Braakman Sluice.

Table 7.12. General data of the Braakman models.

|  | Detailed model | Generalized model |
|---|---|---|
| Areas and Boundaries | 11 | 6 |
| Subsystems simulation | 25 | 12 |
| Subsystems optimization | - | 7 |
| Regulating structures | 5 | 4 |
| Fixed structures | 5 | 1 |

*Current Practice*

Runoff has been verified, using the detailed model of the Braakman water system for the average hydrological year of 1992. Only a general water balance could be made, because no data on the Braakman Sluice discharges and only general operational data on the Isabella

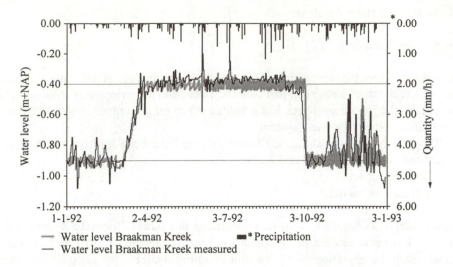

*Fig. 7.30. Water levels in the Braakman Kreek: measured (manual control) and calculated (local automatic control) for the average hydrological year of 1992.*

Pumping Station are available. Sluice discharges have been verified roughly using water-level measurements from the Braakman Kreek and the Westerschelde and by a general model of the sluice structure (Sec. 4.3.4 and Janssens, 1994).

Figure 7.30 shows the fluctuation in water level in the Braakman Kreek, based on measurements and according to the detailed model calculations assuming local automatic control of all regulating structures. Despite the fact that the two curves are presented in one figure, it should be kept in mind that water levels obtained are incomparable because the modes of control are completely different: manual control of both the Isabella Pumping Station and the Braakman Sluice, versus local automatic control of the two structures.

Several interesting characteristics can be observed in the figure:
- the strong fluctuations in water levels;
- the gentle rise in water level in March;
- the sudden fall in water level at the end of September.

The water-level fluctuations are caused by discharges of the Isabella Pumping Station to the Braakman Kreek, especially during periods when discharge via the Braakman Sluice is not possible. A second cause is the manual control of the Braakman Sluice sliding gates. Some over-reactions in operating these gates can be observed; immediately after a high water level, the gates were sometimes opened for too long. This resulted in too low water levels. In the model, the four sliding gates are controlled locally, which enables the target level to be maintained better.

In November and December, the water level in the Braakman Kreek started to rise even more, while no extreme precipitation occurred. This rise in water level was the result of too high water levels in the Westerschelde river, which restricted gravity discharge (see

Fig. 7.27 for details). The figure shows very clearly, that despite the efforts of the Water Board in trying to maintain the winter target level, the summer target level, which is 0.5 m higher, was almost reached.

The gentle rise in water-level in March is the result of switching from winter to summer target water levels. During that period the Braakman Sluice was kept almost entirely closed, to preserve water.

The sudden fall in water level at the end of September is the result of switching from summer to winter target levels.

When judging the modeled water levels, it should be kept in mind that the data used for verification are very general. Precipitation for instance, was measured only at one location and available as a daily total. The measured precipitation is included in the calculation, assuming that it was the same over the entire water-system area. It is well-known however, that local precipitation measurements are generally not entirely representative for large areas, especially not in summer. For this reason, it is very likely that while several local precipitation events were monitored, others may have been missed.

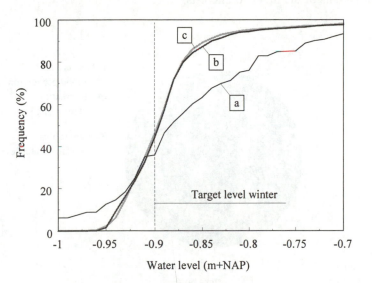

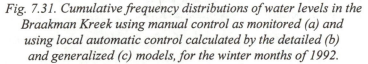

*Fig. 7.31. Cumulative frequency distributions of water levels in the Braakman Kreek using manual control as monitored (a) and using local automatic control calculated by the detailed (b) and generalized (c) models, for the winter months of 1992.*

Figure 7.31 shows the cumulative frequency distribution of observed and calculated water levels during the winter months of 1992. The periods used are from 1-1-92 to 28-2-92 and from 1-10-92 to 31-12-92. The figure shows a difference between the water level measured and the water level calculated, using local automatic control. Extremely high water levels occur less frequently, when local automatic control is used than when manual control is used.

The better water level control is the result of closing the gates at the closing setpoint of NAP - 0.95 m, using local automatic control. The target water level can be maintained more accurately by opening the gates to the required level, so that enough water is discharged from the Braakman Kreek, but not too much.

### Interests Considered

In this case study only the interests in the controllable part of the water system are considered: the common-good interest flood prevention, the sectoral interest recreation and operational water-management interests.

Figure 7.32 shows the various interests distinguished and how these interests are mutually weighed. The flood-prevention interest together with the recreational interest receive a relatively large weight. The operational interests receive a relatively small weight. This means that the requirements of the latter interests can be violated when necessary to satisfy the requirements of the first two interests.

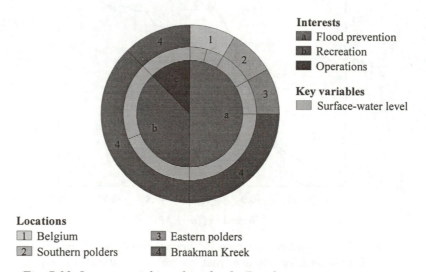

**Interests**
- a Flood prevention
- b Recreation
- Operations

**Key variables**
- Surface-water level

**Locations**
- 1 Belgium
- 2 Southern polders
- 3 Eastern polders
- 4 Braakman Kreek

*Fig. 7.32. Interest-weighing chart for the Braakman water system.*

Figure 7.33 shows schematic damage functions as used in the generalized model. Recreational interests are met optimally in summer when the surface-water level is at NAP - 0.4 m, with a preferred range of 0.10 m. The damage function for the operational interest represents that situation in which no damage to embankments along the Braakman Kreek occurs.

In the winter season the flood prevention interest prevails. The safety margin for this interest is given by the level which should not be exceeded for the Braakman Kreek to maintain enough storage capacity to cover two tidal periods during which Isabella Pumping Station is at full capacity, while the Braakman Sluice cannot discharge. This damage function implicitly includes the situation that the Isabella Pumping Station is controlled manually.

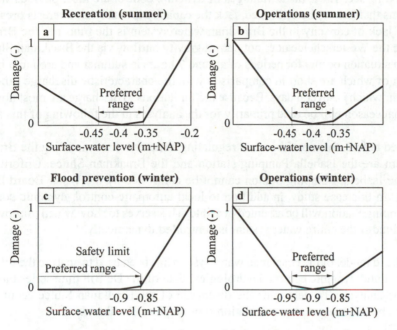

Fig. 7.33. Example shapes of damage functions used.

### Control Alternatives

The following control alternatives are distinguished in assessing the performance of the Braakman water system:
- manual control,
- local automatic control,
- dynamic automatic control.

The current practice is remote manual control. The manually controlled gates and the difference in water level between the Braakman Kreek and the Westerschelde determine the discharge of the sluice.

Local automatic control involves automatic operation of the sliding gates on the basis of water levels in the Braakman Kreek and the Westerschelde. Two conditions have to be fulfilled to open the gates: the water level of the Braakman Kreek should be above its target level and the water level of the Westerschelde should be below that of the Braakman Kreek.

Dynamic control involves remote control of the sliding gates. The following conditions have to be fulfilled before the sliding gates are opened: discharge is needed to satisfy the combination of weighed interests and; the water level of the Westerschelde should be below that of the Braakman Kreek.

As explained in Sec. 1.3.1, the advantages of dynamic control are most obvious when the water systems that have to be controlled, lack the capacity to satisfy all interests present. An example of lack of capacity in the Braakman water system is the runoff to the Braakman Kreek while the Westerschelde does not permit gravity outflow via the Braakman Sluice. In general, this situation occurs for periods of around 7 hours in summer and around 5 hours in winter, both of which are short in comparison with the characteristic discharge time of the water system, which is several days. Because the limitations to discharge are most noticeable in winter, that season will be used primarily for the analyses in the following of this section.

As mentioned before, the most important regulating structures for controlling the Braakman water system are the Isabella Pumping station and the Braakman Sluice. Unfortunately, currently the Isabella Pumping Station cannot be controlled by the Water Board De Drie Ambachten. In this case study, in addition to local automatic control, dynamic control of Isabella Pumping Station will be assumed possible. This serves to show which improvements can be obtained if the entire water system is controlled dynamically.

The generalized model of the Braakman water system has been used to analyze the difference between local and dynamic control. Hydrological data of the De Bilt time series have been used for this analysis. To determine the discharge of the Braakman Sluice accurately, it turned out to be necessary to use calculation time steps of 30 minutes.

The winter season 1974/1975 was chosen for the analysis, because in this period precipitation was relatively high, especially in the months October to January (October: 140 mm, November: 121 mm, December: 124 mm, January: 88 mm). An uncertainty multiplier of unity has been used in the prediction. The effect of incorrect predictions is analyzed separately in the following sections. The following prediction scheme has been used as a basis for the calculations:
•      0 to 3 hours: perfect prediction;
•      3 to 6 hours: period-average prediction;
•      6 to 12 hours: period-average prediction;
•      12 to 18 hours: period-average prediction.

The control horizon has been chosen to include two periods of low tide, e.g. 18 hours. An optimization time-step size of 30 minutes has been used, to accurately incorporate water level differences in the few hours available for discharge from the Braakman Kreek into the Westerschelde.

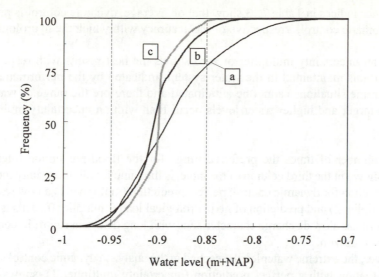

*Fig. 7.34. Cumulative frequency distributions of the calculated water
level in the Braakman Kreek, using local automatic control (a) and
dynamic control, excluding (b) and including (c) the Isabella
Pumping Station, for the winter season of 1974/1975.*

Figure 7.34 indicates an improvement in maintaining the water level of the Braakman Kreek when, instead of local automatic control (a), dynamic control of the regulating structures is applied, excluding the Isabella Pumping Station (b). The improvement is once more of the same order of magnitude as the improvement caused by switching from manual to local automatic control (compare Fig. 7.31). Curve (c) in Fig. 7.34, shows a further improvement of results. It presents the situation in which Isabella Pumping Station is also dynamically controlled.

The associated performance indices for the curves in Fig. 7.31 are: 1 for local control (a), 1.8 for dynamic control excluding the Isabella Pumping Station (b) and 2.3 for dynamic control including Isabella pumping station (c).

The accuracy of the predicted hydrological load as indicated above, has been further examined, by simulating incorrect weather forecasts. Table 7.13 lists the results of dynamic control, including Isabella Pumping Station, using an uncertainty multiplier of two, in fact, overestimating the hydrological load by 100% and using an uncertainty multiplier of zero, in fact, incorporating no hydrological load at all.

In addition, the performance index for the water system during the period calculated has been determined, as well as the number of times the flood-prevention interests were failed.

The performance indices in Table 7.13 show that on average, dynamic control is preferable over local automatic control, irrespective of the accuracy with which the hydrological load is predicted.

Using an uncertainty multiplier of zero shows the best results with respect to the average conditions maintained in the water system, indicated by the performance index. However, extreme situations cannot be anticipated and therefore the range of water-level fluctuation is larger and higher water levels occur than with an uncertainty multiplier of unity.

The relative number of times the preferred range for the flood-prevention interest was exceeded, as shown in the third column of the table, is the same for all uncertainty multipliers applied in prediction for dynamic control: perfect prediction (multiplier 1) an overestimation by 100% (multiplier 2) and prediction of no hydrological load (multiplier 0). This is because the water system has slow discharge characteristics with long delays in runoff in comparison with the control horizon.

However, the extreme water levels reached in the analyzed dynamic control situations differ. The situation with a perfect prediction (uncertainty multiplier 1), shows the best results: the range of fluctuation throughout the winter season is 0.16 m, which is clearly smaller than the ranges with prediction error and no prediction.

During extreme events, a good prediction of the water-system load prevents dynamic control taking incorrect control actions. Examples of such incorrect actions for the Braakman Sluice are in case of an overestimate: unnecessary discharge of water in advance; or in case of no prediction: preventing discharge to preserve water, while an extreme rainstorm is approaching.

Table 7.13. Extremes analysis for local and dynamic control and sensitivity analysis of system load prediction (winter season 1974/1975).

| Control mode and uncertainty multiplier | Performance index | Duration of interest failures | | | Extreme water levels | | |
|---|---|---|---|---|---|---|---|
| | (-) | < NAP -0.95 m | > NAP -0.85 m | Total | Lowest (m+NAP) | Highest (m+NAP) | Range (m) |
| Local automatic control | 1 | 4% | 23% | 27% | -0.98 | -0.71 | 0.27 |
| Dynamic control uncertainty multiplier 1 | 2.3 | 2% | 1% | 3% | -0.97 | -0.81 | 0.16 |
| Dynamic control uncertainty multiplier 2 | 2.3 | 2% | 1% | 3% | -0.99 | -0.78 | 0.21 |
| Dynamic control uncertainty multiplier 0 | 2.7 | 1% | 2% | 3% | -0.96 | -0.77 | 0.19 |

An analysis similar to the one for the winter season of 1974/1975 has been performed for the summer season of 1992. The results of this analysis show that during that summer, using the basic prediction scheme with an uncertainty multiplier of unity, the performance index for dynamic control becomes 3.5, a considerable improvement in comparison to local control. The performance indices obtained by using uncertainty multipliers of zero and two, prove again that, on average, the availability of an accurate prediction of the hydrological load is not really necessary for the Braakman water system.

Continuously high water levels may occur in practice in the Westerschelde, which does not permit sufficient gravity outflow during low tide. If a prediction of such periods is not available, dynamic control on these occasions is not a suitable option to improve water-level control.

However, once accurate predictions of high water levels in the Westerschelde become available, including the effect of wind, the situation may improve strongly. In that case the anticipating effect of dynamic control will force the Braakman Sluice to discharge water in advance, even before the runoff from the water system has reached the Braakman Kreek.


*Extremes Calculation*

To give an impression of how the dynamic control phenomenon works for the Braakman water system, Fig. 7.35 presents the results of time-series calculations. The runoff in the period of the severe rain storm of July 1965 (Table 5.1, Page 95) has been calculated. In this case the summer season has been chosen because the average surface-water level is already high in summer, which makes the water system sensitive to a further water-level increase as a result of extreme hydrological loads. The groundwater levels just before this rain storm were high as a result of precipitation in the preceding period.

For demonstration purposes, the graphs of Fig. 7.35 present a very short interval of the entire period calculated, which covered 1-1-65 to 19-6-65. Table 7.14 lists the results of the extreme water levels obtained in the calculations.

Local automatic control of the Braakman Sluice and the Isabella Pumping Station (a), shows that there is a time lapse of several hours before the runoff to the surface water starts. In the mean time, the water level in the Westerschelde has risen above the level at which the Braakman Sluice can discharge. For that reason the water level in the Braakman Kreek continues to rise, until the next low tide. The highest water level in the Braakman Kreek during this event is NAP - 0.22 m; far above the upper limit of the preferred range (NAP - 0.35 m).

In the option applying dynamic control of the Braakman Sluice and local automatic control of the Isabella Pumping Station (b) the possibility to discharge water in advance to the Westerschelde is utilized. Water is discharged from the Braakman Kreek, even before the water level has risen above the local setpoint for opening the gates. Discharge from the Belgian Area by the Isabella Pumping Station, however, takes place during the period that no water can be discharged via the Braakman Sluice. For this reason, the water level in the

Braakman Kreek still rises above the upper limit of the preferred range; to NAP - 0.33 m. The water level in the Belgian Area shows exactly the same result when local automatic control of the entire water system is applied.

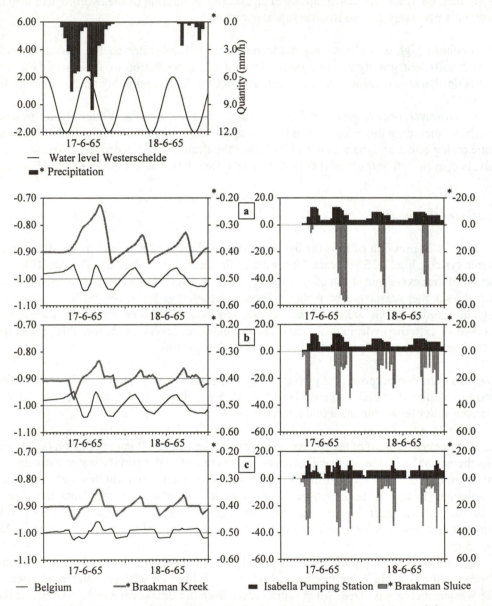

Fig. 7.35. Time-series calculation results for local automatic control (a) and dynamic control, excluding (b) and including (c) Isabella Pumping Station (on the left: levels in m + NAP; on the right: flows in m³/s).

The option of dynamic control of the Braakman Sluice and the Isabella Pumping Station (c) presents the best results. The highest water level, NAP - 0.34 m, is still outside the preferred range, but lower than in the other two options. Moreover, the range of water-level fluctuation is only 0.11 m. In addition to discharging water to the Braakman Kreek in advance, water is especially pumped by the Isabella Pumping Station around the periods that the gates of the Braakman Sluice can be opened. In fact, water is pumped into the Braakman Kreek and shortly afterwards discharged via the Braakman Sluice. The dynamic water-system control option, shows the best results in the Braakman Area as well as in the Belgian Area.

Table 7.14. Calculated water-level extremes and fluctuations as a result of local automatic control (a) and dynamic control excluding (b) and including (c) Isabella Pumping station, in the period 16-6-65 to 19-6-65 (summer season).

|  | Lowest water level (m + NAP) | Highest water level (m + NAP) | Range (m) |
| --- | --- | --- | --- |
| a. Local automatic control | -0.44 | -0.22 | 0.22 |
| b. Dynamic automatic control excluding Isabella Pumping Station | -0.48 | -0.33 | 0.17 |
| c. Dynamic automatic control including Isabella Pumping Station | -0.45 | -0.34 | 0.11 |

Since it is important to use a balanced form of control, the situation in the Belgian Area should not become worse as a result of dynamic control. This has further been investigated for the winter season of 1974/1975. Figure 7.36 shows that the water level in the Belgian Area in that period remains within the preferred range in both control options.

## 7.3.5 Conclusions

### Results

This case study has shown that local automatic control of the Braakman Sluice would satisfy the Braakman Kreek interests considerably better, than the current manual control. This is the result of maintaining the surface-water level better, which is the only controllable variable in the Braakman Area at the moment. In addition, this result can be further improved by dynamic control of the Braakman Sluice. Further enhancements can be obtained by

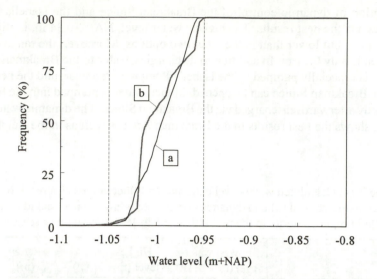

*Fig. 7.36. Cumulative frequency distribution of water levels in the
Leopold Kanaal (Belgium) as a result of local automatic control (a)
and dynamic control (b) of the Isabella Pumping Station, for the
winter season of 1974/1975.*

dynamic control of the Isabella Pumping Station, which is presumably operated on the basis of the water level in the Leopold Kanaal in Belgium.

Considering the currently available system capacities, dynamic control of both Braakman Sluice and Isabella Pumping Station, offer the best chance of improving water-system performance, in the Belgian Area as well. This option requires the Dutch Water Board and the Belgian water authorities to join forces.

When the hydrological-load prediction is not very accurate, it turns out to be better to control the water system dynamically, without hydrological-load predictions. The situation without prediction shows almost the same results as the one with prediction. During extreme precipitation, however, predictions support a more balanced discharge of water via the Braakman Sluice.

In the present-day situation, the Braakman Sluice is not equipped with pumping units. Therefore, when the water levels in the Westerschelde are too high, as a result of a continuing storm, problems in the Braakman Kreek will remain. These problems will increase in the Braakman water systems, if the average sea level rises. In similar systems along more upstream tidal river courses, a rise of the level of the river bed may have a same effect.

The tidal fluctuation of the Westerschelde is known in advance, but at present no accurate prediction of wind effects on the water level is available to the Water Board. Once such predictions are available, dynamic control with an extended control horizon of, for instance, several tidal periods, would be an option to automatically prevent extreme water-level rises during continuous periods of high water levels in the Westerschelde.

Moreover, if the main regulating structures of the water system are controlled together, potential water-level rises in the Braakman Kreek can be reduced if water-level rises in the Leopold Kanaal are accepted. One way to accomplish this, is by reducing the permitted capacity of the Isabella Pumping Station during periods that the Braakman Sluice cannot discharge. If the Isabella Pumping Station cannot be controlled dynamically, such a control strategy can be incorporated in its local control logic.

The results obtained in this case study can be used to determine the potential for dynamic control of other water systems. For instance, the water system east of the Kanaal van Gent-naar-Terneuzen, which includes District 1 and 2 (see Fig. 7.24, Page 201) has a great potential. There the results of dynamic control will probably be even better than those which can be obtained in the Braakman water system. The main reason is that the Otheensche Sluice is equipped with pumping units, which can be operated during periods when gravity discharge to the Westerschelde is not possible. Dynamic control would use these pumping units only if necessary to maintain a good water-system performance. This will be the case if gravity outflow via the sluice is not possible for a long period and/or after high precipitation.

### Further research

The current layout and structures in the Braakman Area limit the options for groundwater-level control to satisfy agricultural and nature interests. To improve control, it would be necessary for local weirs to be automated. The soil conductivity most likely enables water preservation. However, the effect of shallow seepage should first be determined, to estimate the amount of water which can be preserved in dry periods. Automation of weirs may be costly, but probably the only option to limit the inlet of alien water.

Once the water quality in the Leopold Kanaal is known from a monitoring network, dynamic control of the Braakman water system could include the option of combined water-quantity and water-quality control. During a dry summer, the water quality in the Braakman Kreek can be maintained by flushing it with water from the Leopold Kanaal. This option can only be applied, obviously, if that water is of a better quality than that of the Braakman Kreek and, moreover, if sufficient water is available.

## 7.4    Salland Case Study

### 7.4.1  Introduction

The hilly area of Salland is situated in the eastern part of the Netherlands (see Fig. 1.1). The total area covers 50,000 ha. The altitude differences in Salland are up to 11 m. The majority of the area suffers from drought in summer and for that reason water supply works are carried out by Water Board Salland, the responsible authority. Water management in the area relies on approximately 350 weirs and water-supply pumping stations. The water systems of Salland are generally considered the most difficult ones to control in water management, because the highly permeable soils restrict the possibilities of water preservation and control.

The water system described in the present study is the Luttenberg Area, with a total surface of 3,700 ha. The Luttenberg Area is bordered to the south and east by the hills of the Hellendoornsche Berg, to the north by the Overijsselsch Kanaal and to the west by the hills of the Luttenberg (Fig. 7.37).

The Water Board is planning to construct new water-supply pumping stations (1996) in the Luttenberg Area to prevent the surface-water subsystems from running dry in summer. It is essential to determine the need for these supply works and, if they are necessary, the way in which these works can be controlled optimally to best satisfy the requirements of the agriculture and nature interests that are present.

### 7.4.2  Water-System Description

The hills that border the Luttenberg Area in the east and the west have highest altitudes up to NAP + 46 m and NAP + 32 m. The valley between the hills slopes gently from the south to the north, from an altitude of NAP + 10 m to NAP + 6 m over a distance of 9 km. The subsurface of the area almost entirely consists of sand formations. The phreatic aquifer reaches down from the soil surface to NAP - 20 m. At that level impervious clay formations exist.

The main interests in the Luttenberg Area are flood prevention, agriculture and nature. Hardly any residential areas exist in this area. Agricultural activities are found mainly in the valley. The hills are covered by deciduous and evergreen forests.

In the past, flooding frequently occurred in winter and spring. This made farming in spring very difficult. For that reason the Water Board has built extensive drainage works in the sixties to discharge excess water faster from the area. Existing water courses were deepened and widened and new canals were constructed. At present a system of relatively small canals exists, in which fixed weirs have been placed to prevent too high flow velocities.

The permeability of the sandy soils is extremely high: 40 m/d on average (Werkgroep Hydrologisch Onderzoek Overijssel, 1976). This high permeability results in fast infiltration into the groundwater aquifer and shallow seepage below weirs (principle shown in Fig. 1.7, Page 10).

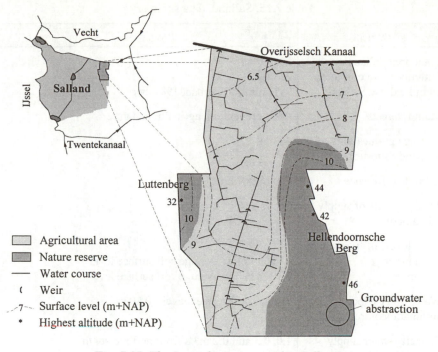

*Fig. 7.37. The Luttenberg Area within Salland.*

After the drainage works were built, the flooding problems in spring were solved, however, groundwater levels have fallen 0.5 to 1 meter on average, during the past thirty years. Water shortages in both the agricultural areas and parts of the nature reserves are one of the main problems in summer. The weirs installed by the Water Board prove not to be suitable to preserve water during dry periods.

For drinking water production, 5.5 million m³ of groundwater is abstracted a year at Nijverdal, on the Hellendoornsche Berg. This amount accounts for approximately 7% of the water balance of the Luttenberg Area. Additionally, farmers commonly use a large quantity of groundwater for irrigation purposes in dry periods.

The above mentioned effects result in such an overall water shortage, that canals are generally dry for more than 100 days a year. Within the framework of the national policy to reduce drought problems, the Water Board has scheduled measures to preserve as much water as possible and, in addition, supply water from the Overijsselsch Kanaal to the area by means of small pumping stations (Salland, 1995). The water of the Overijsselsch Kanaal originates from the IJssel (via the Twentekanaal) and the river Vecht (Fig. 7.37).

In the remainder of this section the water system of the Luttenberg Area will be discussed, which therefore is called the Luttenberg water system. Details of the water system are listed in Table 7.15. Further details can be found in Schuilenburg (1996).

Table 7.15. Salland, data summary.

| Subject | Properties [1] |
|---------|----------------|
| Total area Salland | 50,000 ha |
| Luttenberg water system | 3,700 ha |
| main land use Salland | 94% rural, 5% urban, 1% other |
| Main interests | flood prevention, agriculture, nature |
| Average permeability of the sandy subsurface | 40 m/d |
| Altitude difference | 4 m over 9 km |
| Total capacity of supply scheduled | 1 m$^3$/s |
| Crest levels of fixed weirs | |
| summer | 0.90 m below average soil surface level |
| winter | 1.10 m below average soil surface level |
| Water level setpoint automatic control of weirs | average soil surface level - 0.80 m |
| Capacity water supply works (specifically for the Luttenberg water system) | 1.0, 0.5 and 0.2 m3/s from north to south |

[1] Luttenberg Area, if not specified

## 7.4.3 Water Management

### Current Practice

The only means of controlling the Luttenberg water system is manual adjustment of weirs at the moment. Manual adjustments are carried out by the Water Board in spring and autumn. During summer, the weirs are used to preserve the maximum quantity of water possible, while in winter they are lowered to permit higher discharges and prevent flooding after excessive precipitation.

Weir crests are generally kept approximately 0.3 m below the lowest soil-surface level in summer and 0.5 m or more below the soil surface in winter. Surface-water quality in the area and in the supplying systems generally meets the standards and is thus not considered an issue in control of the water system by the Water Board at the moment.

*Problem Formulation*

The groundwater table in the Luttenberg water system has fallen considerably as a result of intensive drainage works, which have been carried out the past few decades. This causes drought problems in the agricultural areas and nature reserves. The Water Board intends to construct water-supply works, e.g. pumping stations, to restore a higher groundwater table and better satisfy the requirements of agriculture and nature interests.

The Water Board has chosen to construct a combination of fixed weirs and locally controlled pumping stations. This option will require the crest levels of weirs to be the same height as is current practice, mainly determined by extreme precipitation, which should not result in flooding of agricultural lands nor areas along water courses in general.

*Aim of the Case Study*

This case study aims at finding the best control mode for the Luttenberg water system, trying to meet the following objectives:
- to prevent flooding along water courses;
- to prevent flooding in agricultural areas;
- to maintain water in surface-water subsystems for irrigation;
- to preserve location-specific water and minimize external water supply;
- to raise groundwater levels in agriculture areas;
- to raise groundwater levels in nature reserves.

Within the scope of this thesis, the main focus is to find the best control solution meeting the flood-prevention, agricultural and nature interests. In the options studied, the present fixed weirs are also assumed to be controllable, which is not planned by the Water Board at present.

## 7.4.4 Water-System Analysis

*Modeling*

Modeling and associated calibration of the Luttenberg Area has been studied and reported by Schuilenburg (1996). Fig. 7.38 shows a schematic representation of the water system, including the supply pumping stations as currently planned.

The natural systems of the hills that border the water system are schematized by groundwater subsystems. These areas are called the Luttenberg, the Hellendoornsche Berg South and the Hellendoornsche Berg North. No water courses exist in these hilly areas and therefore groundwater is the only means of flow interaction with other subsystems. This results in a groundwater submodel (indicated by light gray flow elements in Fig. 7.38). The groundwater abstraction for drinking-water production is located on top of the Hellendoornsche Berg South, at the border of the water system modeled.

Runoff from the hills flows to the southern and to the central and northern part of the valley and therefore these areas are called: Southern Area, Central Area and Northern Area. In practice, several weirs exist in the water courses in these areas. These have been schematically represented by three weirs. To incorporate the effect of shallow seepage, groundwater flow elements connecting up- and downstream surface-water subsystems are modeled parallel to the weirs.

Similar to the weirs, the actual groundwater flow pattern is more complex with seepage and infiltration sections inside the areas distinguished. However, for the present water-control study, a schematization into several larger groundwater flow elements is considered sufficiently accurate.

Excess water leaves the water system in the north, and flows into the Overijsselsch Kanaal. The influence of the groundwater potential at the other side of this canal is incorporated in the model through the boundary called Groundwater North.

Fluctuating inflows and outflows during the summer and winter seasons in the south are also modeled by boundaries, called Groundwater South and Groundwater Hellendoornsche Berg.

In the water-supply alternatives studied, water is pumped up in three stages from the Overijsselsch Kanaal, via the Northern Area and the Central Area to the Southern Area. In each of these areas, water is used or flows away via the groundwater and therefore the pumping capacities decrease from north to south.

Table 7.16 presents the subsystems and flow elements distinguished in the model, including the groundwater flow elements, together forming the groundwater submodel.

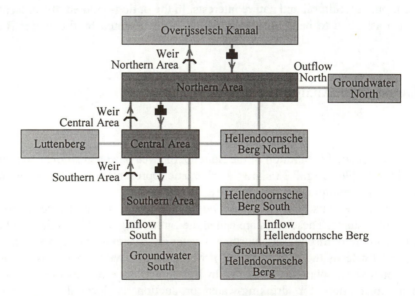

*Fig. 7.38. Schematic representation of the Luttenberg water system, showing the situation after constructing the water-supply works.*

Table 7.16. General data of the Luttenberg model.

|  | Number |
|---|---|
| Areas and Boundaries | 10 |
| Subsystems simulation | 13 |
| Subsystems optimization | 9 |
| Regulating structures [1] | 0 / 3 / 6 |
| Fixed structures [1] | 3 / 3 / 0 |
| Groundwater flow elements | 10 |

[1] Alternatives: fixed weirs only / fixed weirs and pumping stations / controlled weirs and pumping stations.

In modeling a sloping area such as the Luttenberg water system by subsystems, special attention should be paid to incorporate the effects of surface slopes and slopes of water courses. In the modeling presented in Chapters 4 and 6, surface-water and groundwater levels are assumed horizontal.

Figure 7.39 presents the principle of converting the sloping properties into flat ones. In the model, average conditions of soil elevation and the canal bottom level are used. Average levels of surface-water and groundwater subsystems are calculated. When interpreting the results, it should be noted that the water levels at the highest and lowest locations within each area generally determine whether drought or flooding occurs.

Especially groundwater levels may show considerable variations within one area. If the shape of the groundwater table has to be calculated more accurately, a numerical

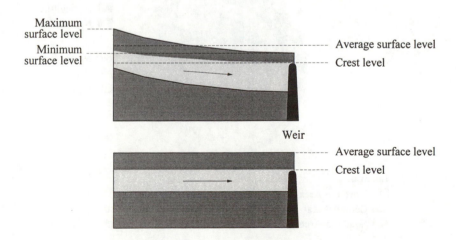

*Fig. 7.39. The principle of modeling a sloping area.*

groundwater model could be made which uses the surface-water levels calculated by the simultaneous simulation and optimization method as input.

### Interests Distinguished

Three interests are distinguished in the Luttenberg water system: flood prevention, agriculture and nature. Figure 7.40 shows that these interests have been assigned equal weights. The interests are furthermore related to the locations in the water system by means of the following key variables:

- Northern Area: surface-water and groundwater level (flood-prevention and agricultural interests);
- Central Area: surface-water and groundwater level (flood-prevention and agricultural interests);
- Southern Area: surface-water and groundwater level (flood-prevention and agricultural interests);
- Luttenberg: groundwater level (nature interest);
- Hellendoornsche Berg North: groundwater level (nature interest);
- Hellendoornsche Berg South: groundwater level (nature interest).

For the agricultural interests, a slight preference is given to groundwater control above surface-water control. This can be observed in the chart by the hatched sectors in the variable ring. The groundwater-related sectors are a little larger than the surface-water-related ones.

Only the water requirements of the natural values of the fringes of the nature reserves can be influenced by means of control. This is because the groundwater table more up hill

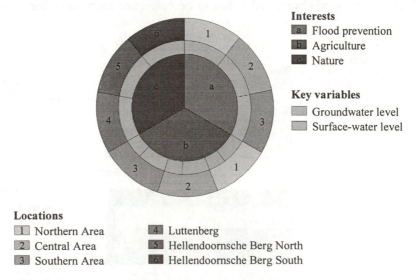

*Fig. 7.40. Interest-weighing chart for the Luttenberg water system.*

is so far below the soil surface that a separate and disconnected water subsystem develops in the topsoil of these hills.

Flooding of the fringes of the nature reserves on the hills cannot occur even during extreme hydrological conditions, because these fringes are relatively high compared with the agricultural areas in the valley (i.e. the Northern, Central and Southern Area)

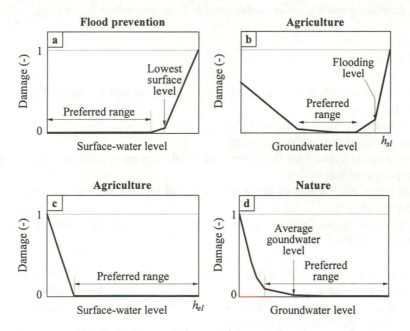

Fig. 7.41. General shapes of damage functions used.

Figure 7.41 shows the general shapes of the damage functions used in the model. Function (a) represents the damage to the flood-prevention interest. Damage to this interest occurs when the surface-water level is above the lowest location in an area and reaches its maximum when the entire area is flooded. The water system is considered to fail as soon as the land along the water courses floods.

Function (b) represents the damage to the groundwater-related agricultural interest. The graph shows that a preferred range of the groundwater level can be distinguished. Within this range the growing conditions for the crop are optimal and no to hardly any damage occurs. The range of approximately 0.5 m is determined by the lowest optimal groundwater level in the high locations and the highest optimal groundwater level in the low locations.

Function (c) represents the damage to surface-water related agricultural interest. The starting point for the introduction of this function is the option of irrigation with surface water in dry periods, avoiding irrigation with groundwater, when the groundwater table (function b) becomes too low. A range which prefers at least 0.2 m of water in the water courses is used. Functions (b) and (c) are both related to the agricultural interest, which means they are mutually weighed (Fig. 7.40). The maximum damage to the agricultural interest occurs when the groundwater level has fallen below the preferred range and the surface-water level has fallen below the level at which water can be abstracted from the surface water for irrigation.

Function (d) represents the damage to the groundwater-related nature interest. The average groundwater level as determined for the original situation is indicated in the figure. The damage to this interest increases with a lower groundwater level, eventually reaching the maximum damage when the groundwater level falls meters below the average.


### Control Alternatives

Given the present layout, five alternatives can be distinguished to control the Luttenberg water system:
1.   local manual control of the weirs;
2.   local automatic control of the weirs;
3.   local manual control of the weirs and local automatic control of the supplying pumping stations;
4.   local automatic control of the weirs and local automatic control of the supplying pumping stations;
5.   dynamic automatic control of the weirs and dynamic automatic control of the pumping stations.

The first alternative represents the current situation with fixed weirs, of which the height is adjusted at turn of the season, i.e. on the first of April and the first of October.

The second alternative represents the situation in which as much water is preserved as possible, using upstream controlled weirs.

The third alternative represents the situation currently planned by the Water Board, comprising fixed weirs and downstream-controlled pumping stations to supply water whenever needed. This alternative implies surface-water setpoints which are 0.2 m higher for the summer season than the winter setpoints.

The fourth alternative is similar to the third, but adds controllable weirs to it, using fixed setpoints for the surface-water level upstream the weirs.

The fifth alternative involves dynamic automatic control of both the weirs and pumping stations. In this alternative the interests as weighed and the conditions in the water system determine the control of regulating structures.

Local automatic control is executed exclusively on the basis of surface-water levels. Dynamic control is based on interest requirements, the satisfaction of which is determined by both surface-water and groundwater levels. No combinations of dynamic automatic and local automatic control are considered, since important information on groundwater levels, used by dynamic control, is not available in local control. This could result in conflicting control actions for weirs and pumping stations during exceptional water-system conditions.

The control horizon applied in dynamic control is 48 hours and comprises perfect, average and scenario types of prediction, which together form the standard prediction scheme used:

- 0 - 8 hours: perfect prediction, uncertainty multiplier 0.6 for precipitation and 0.8 for evaporation;
- 8 - 24 hours: period-average, uncertainty multiplier 0.5 for precipitation and 0.8 for evaporation;
- 24 - 48 hours: period-average, uncertainty multiplier 0.5 for precipitation and 0.8 for evaporation.

Using the uncertainty multipliers means underestimating the extremes in both precipitation (by 40% and 50%) and evaporation (by 20%). For that reason control strategies determined by dynamic control will take into account less extreme conditions and generally determine gentle control actions. In this way, underestimates are incorporated, which would occur in forecasting by a weather bureau.

## Water-System Performance

Time-series calculations have been executed using the model of the Luttenberg water system and using the 30-year time series of De Bilt. Table 7.17 presents the results obtained and discussed below. The setpoints for the regulating structures of the various control alternatives have been determined and were set optimally for the various control alternatives. This has been done to allow a well-balanced comparison between the control alternatives, all setpoints were determined and set so that flooding along water courses and flooding of agricultural land occurred rarely: less than 0.1 a year and less than 0.5 a year respectively.

Table 7.17. Results of time-series calculations for five control alternatives of the Luttenberg water system.

| Alternative | | Performance index (-) | Failure frequency agriculture interest (d/y) | Preservation factor (-) | rise in gwl valley [1] (m) | rise in gwl hills [2] (m) |
|---|---|---|---|---|---|---|
| 1. Weirs | LM | 1 | 223 | 0.67 | - | - |
| PSs | - | | | | | |
| 2. Weirs | LA | 1 | 214 | 0.91 | 0.05 | 0.05 |
| PSs | - | | | | | |
| 3. Weirs | LM | 13.1 / 1 | 0 | 0.6 | 0.54 | 0.71 |
| PSs | LA | | | | | |
| 4. Weirs | LA | 13.5 / 1.03 | 0 | 0.7 | 0.6 | 0.76 |
| PSs | LA | | | | | |
| 5. Weirs | DA | 20.2 / 1.54 | 0 | 0.66 | 0.7 | 0.86 |
| PSs | DA | | | | | |

[1]  Average rise in groundwater level valley, relative to alternative 1
[2]  Average rise in groundwater level hills, relative to alternative 1

The current situation in the water system, alternative 1, has been used to determine the long-term evaluation parameter of the performance index. This alternative comprises three fixed weirs in the main water courses. In the higher parts of the water system, water courses are normally dry for more than 170 days a year on average (not in the table), however, irrigation is not possible during 223 days a year.

Alternative 2, in which the weirs are automated, shows no improvement in performance in the long term. The main reason is that automatically controlled weirs can only reduce the water-system outflow for very short periods: days to weeks at most, while the water courses remain dry for very long periods. For that reason the agricultural interest is still violated 214 days a year using this alternative.

     Alternative 2 preserves considerably more water than alternative 1. The preservation factor is determined by Eq. 7.3, using the total quantity abstracted from the system. This quantity is the sum of evapotranspiration, outflow via groundwater to adjacent water systems and outflow over Weir Northern Area. The preservation factor has increased from 0.67 to 0.91. This effectiveness results from the fixed weirs having lower crest levels than the setpoints of the controlled weirs and the controlled weirs preventing unnecessary outflow when the upstream water level falls below the setpoint. The average groundwater levels in both the agricultural areas and the nature reserves rise 0.05 m as a result of installing upstream-controlled weirs.

In alternative 3, the weirs are again fixed, but supply pumping stations are added, which are automatically controlled on the basis of their downstream water-level setpoints. In this alternative, the performance index makes a huge jump from 1 to 13.1. This is mainly due to the agricultural interests not being violated anymore.

     In general, enormous improvement in performance can be expected when comparing almost uncontrolled water systems with fully controlled ones. To compensate for this effect in the following comparisons, the performance index has again been reset to unity.

     The pumping stations supply large quantities of water to the Luttenberg water system, trying to maintain high surface-water levels and corresponding higher groundwater levels. As can be read from the last two columns of the table, the average groundwater levels in the agricultural areas and nature reserves rise by 0.54 and 0.71 m respectively. However, it should be noted that the actual rise may vary significantly within an area. The result obtained is important in view of better satisfying the interest requirements of agriculture and nature. However, the disadvantage is that the natural groundwater storage capacity in the soil, which develops in summer using alternatives 1 and 2, is considerably reduced in alternative 3. For that reason during precipitation, more water is discharged from the water system than in alternatives 1 and 2.

     The preservation factor is determined irrespective of the source of the water flowing out of the water system and therefore also unused water supplied by pumping is incorporated. For that reason the preservation factor decreases to 0.60.

An added advantage of alternative 4 over alternative 3 is that the weirs are upstream-controlled. For that reason the surface water-level can be kept higher on average. Furthermore, the average groundwater levels in the agricultural areas and the nature reserves rise by 0.60 and 0.76 m. This results in an increase in performance by a factor of 1.03 in comparison to alternative 3. The automatically controlled weirs can preserve surface water better (factor 0.7) in this alternative than in alternative 3 which uses fixed weirs.

Alternative 5 dynamically controls the three weirs and the three supply pumping stations. During the time-series calculations, dynamic control determines its strategy on the current situation in the water system and a prediction of hydrological load. The prediction scheme discussed before has been used. The performance index of this alternative in comparison to alterative 3, increases by 1.54. This result is primarily due to better satisfying groundwater-related requirements in times of drought. Consequently, the groundwater levels in the agricultural areas and nature reserves rise to 0.70 and 0.86 m on average.

Despite the improvements in performance and groundwater-level rise, the preservation factor of 0.66 is slightly lower than in alternative 4. This is due to the larger quantity of water required to supply the groundwater subsystems. It should be kept in mind that, if no water is used in the water system, the rising groundwater also supplies the surrounding, lower-lying areas, with water. These outflows cannot be kept in the water system. Therefore, in this water system, all alternatives that further improve on the rise of the groundwater table, will have a lower preservation factor.

### Comparing Control Modes

To demonstrate the general phenomenon of control in hilly areas, the three types of control for the Central Area of the Luttenberg water system are further analyzed, each in combination with downstream-controlled supply pumping stations:
- using manual control with a fixed weir;
- using local automatic control with an upstream-controlled weir;
- using dynamic automatic control of the weir.

Figure 7.42 shows the results of three control modes in a wet fortnight during the winter of 1980, in which 74 mm of precipitation fell. The period has been selected because during the various time-series calculations with the three modes, the conditions at 29-02-1980 appeared to be very similar, which makes the comparison of results realistic. As in the time-series calculations described before, the standard prediction scheme has been used.

In graphs (a) of Fig. 7.42, the weir crest is fixed at NAP + 6.90 m. Because the upstream water level in the Central Area is continuously above the weir crest, an outflow from that area to the Northern Area results for the entire period. The groundwater level is not controlled explicitly and therefore it rises above the preferred range, which is up to NAP + 7.30 m.

In graphs (b) the weir crest is controlled on the basis of the water level in the upstream water course. The right graph clearly shows that the PI controller used for upstream control is very well able to maintain a stable upstream water level. However, using a fixed winter setpoint of NAP + 7.20 m does not prevent the groundwater table from rising clearly above the upper limit of the preferred range of NAP + 7.30 m.

Graphs (c) show the result of dynamic control of the water system. Here, not only the surface-water requirements, but also groundwater requirements following from the interests play a role. In this specific case, the requirements of the agricultural interest, related to the groundwater subsystem, are violated. To minimize the damage to that interest, the weir is lowered 0.55 m in three days. This effectively results in a fall in surface-water level by 0.20 m in two days and a subsequent outflow from the soil and fall in groundwater level. The

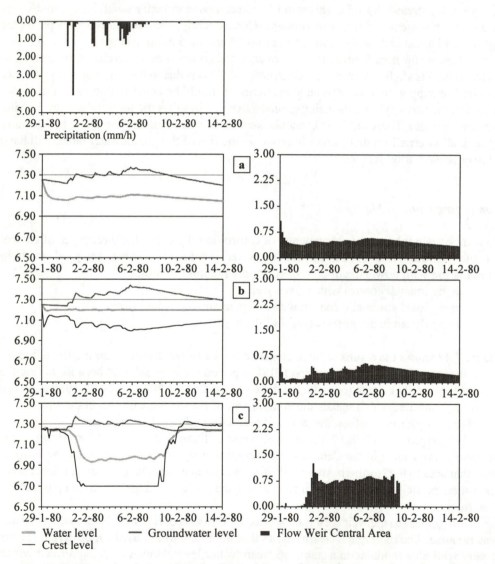

Fig. 7.42. Water-level control in the Central Area of Luttenberg, resulting from three control modes: fixed weir (a), local upstream-controlled weir (b) and dynamically controlled weir (c) (on the left: water levels in m+NAP; on the right: flow in m³/s).

control of the weir in the Central Area thus results in a lowering of the groundwater table by 0.10 m in comparison with local automatic control (b).

A first glance at the graphs shows no advantage of dynamic control with respect to maintaining the groundwater table within the preferred range, in comparison to manual control with a fixed weir (a). However, it should be noted that the weir is fixed at a relatively low level: NAP + 6.90 m, to prevent flooding under extreme conditions. Looking at the end of the left graphs (a) and (c), it becomes clear that the fixed weir cannot prevent the groundwater table from falling. However, this is successfully achieved by dynamic control (c) and within the visible frame also by local automatic control (b).

The water balance (not shown) indicates that the discharge from the Central Area in dynamic control mode is limited to approximately the total volume of precipitation in the period shown.

## 7.4.5  Conclusions

This case study shows that in hilly areas with highly permeable soils such as in the Luttenberg Area, control of surface-water and groundwater levels by fixed weirs only is virtually impossible. Water courses fall dry in summer during more than 100 days, which means that to satisfy the requirements of agricultural interests, farmers have to abstract groundwater for irrigation, which, added to the natural lowering as results from outflow, results in an additional fall of the groundwater table. Combined with the situation that relatively large quantities of water are abstracted from the groundwater for drinking-water production, water boards such as the Salland Water Board experience difficulties with regard to maintaining optimal conditions for agriculture and nature interests.

In principle, three main options can be distinguished to solve these problems in the Luttenberg Area:

1.      Preserve the maximum of water in the area, by closing the drainage systems. This option obviously leads to frequent flooding in winter and therefore the lands will become unsuitable for agricultural purposes. However, natural values can again develop.

2.      Restrict surface water outflow by making water courses less deep. By doing so, the groundwater outflow to the surface-water subsystems is restricted and so is the possibility for surface water to discharge from the area. To maintain the required discharge capacity, water courses could be widened.

3.      Keep the drainage system as it is and construct water-supply works. This will allow both agriculture and nature interests to be satisfied better. However, relatively large quantities of alien water will have to be let into the water system.

In the present case study the third option has been investigated. Building water-supply works in the water system has the advantage that surface water is available whenever required.

If water supply works are combined with fixed weirs, groundwater levels in the valley rise by 0.54 on average. In hilly areas the average groundwater levels rise by 0.71 m. This option has been selected by the Water Board and on the basis of the results, it can be concluded that the fall in water level which resulted from the intensive drainage of the area the past few decades, can be restored to a large extent.

Even better results can be obtained by dynamic control of the supply works and the main weirs in the water system. The groundwater level in the sloping area rises by 0.70 m and in the hilly areas by 0.86 m on average.

The additional rise in groundwater table by approximately 0.15 m, when dynamic control is applied, results from the fact that this control mode uses both surface and groundwater variables to determine the best control strategy. Dynamic control continuously tries to keep the groundwater level as high as possible, but takes preventive measures to obtain the required discharge capacity in water courses if required, e.g. by automatically lowering the weirs. By doing so, dynamic control prevents flooding situations in agricultural areas.

Considering the objectives set for control of the water system, it can be concluded that alternative 3, including locally controlled water supply works and fixed weirs, presents very good results. With respect to the rising groundwater table, stronger improvements can be obtained by dynamic control of water supply works and weirs. However, water is slightly less well preserved in this alternative, since the high groundwater tables also increase the groundwater outflow from this water system to adjacent water systems.

## 7.5    Performance of the Method

### 7.5.1  General

This section describes the performance of the simultaneous simulation and optimization method developed and described in Chapter 6. The method has been implemented in the Decision Support System AQUARIUS, of which the results will be presented. The Strategy Resolver of AQUARIUS consists of two submodules which have been introduced as the Constraint Manager and the Solver.

The Constraint Manager basically builds the optimization problem, by converting the various equations of continuity, damage functions and intrinsic water system relationships into constraints. The structure of that problem is built at the beginning of each time-series calculation. The optimization problem is defined in LP format, by a matrix in which the rows represent the constraints and the columns represent the water-system variables modeled. During the successive simulation and optimization runs, the coefficients in the matrix are updated. This is done for the line segments of damage functions, which may change as a result of time-varying interest requirements; for continuity equations, which include the flows via regulating structures and which can be available or not; and for the side constraints, which describe the linearized flow processes via flow elements.

The matrix coefficients of linearizations are frequently updated and fully determined by forward estimating. In the total updating process, determining the linearization constants is expected to be time-consuming and therefore it is evaluated separately below.

The solver implemented in AQUARIUS is BPMPD, which uses a primal dual interior-point method (Mészáros, 1996). In a comparative study, this solver proved to be the most efficient one available to include in the DSS (Hoogendoorn, 1996). BPMPD communicates with the optimization module of AQUARIUS via a Dynamic Link Library, which makes communication most efficient. However, other solvers can be incorporated in the DSS as well, using communication via files. For that type of communication, the Constraint Manager automatically builds files in MPS format.

The BPMPD solver can be used in two different ways: it can start its search procedure from scratch, or it can start its search from the optimal solution found in an earlier iteration of the SLP method. The latter has been introduced as a warm start. The results of the warm-start option depend strongly on whether the solution found in the earlier iteration, is close to the optimum for the current iteration. To prevent these uncertainties from influencing the results, the warm-start option has not been used in the analyses described, but is separately evaluated.

For analysis purposes, generalized models were used in the case studies, which will be called 'basic models' here. Geographical names are used for the basic models: the Delfland model (Delfland case study), the Braakman model (De Drie Ambachten case study) and the Luttenberg model (Salland case study).

Calculating time series for the basic models, took 1 to 3 seconds for each simultaneous simulation and optimization loop. In the approach followed, time series for 30 years were calculated, using simulation time steps of 0.5 and 1 hour and optimization time steps of 0.5 to 4 hours. It took several hours to days to calculate an entire period of 30 years, using these time steps.

In real-time control, the control strategy may require more detail than obtained in the case studies. Therefore, a special analysis has been performed to determine the applicability of the DSS with respect to solution-time requirements. In real time, the intervals between the successive runs of the Strategy Resolver will be in the order of magnitude of minutes for very fast-reacting systems, to hours for slow-reacting ones. The various analyses proved that, irrespective of the size of the model built, the time needed for simulation was negligible in comparison with the time needed for optimization (less than 1% of the total time per loop). For that reason, only the time needed for optimization is discussed in the following.

A practical limit of 10 minutes has been chosen as the maximum time allowed for the Strategy Resolver to determine a control strategy. This can also be considered a practical limit for operators who are waiting for a reply from the Strategy Resolver during their work and during training.

## 7.5.2 Approach

The models built for the three practical case studies have been extended to create larger problems. Model extensions have been included in the following ways:
- replication of the water-system model and simultaneous running of separate models;
- extending the number of time steps in the control horizon;
- extending the replicated models by connecting them via extra flow elements.

The first two options can be expected to yield comparable results, since the Constraint Manager replicates the entire model structure from the basic nodes into time nodes (Fig. 6.8, Page 118). This is done as many times as optimization time steps fit in the control horizon. It can be expected that despite the small differences in the matrix structure, finding the solution to an optimization problem of a model which is exactly twice the size of the basic model, yields a performance similar to the single model which uses twice as many optimization time steps in the control horizon.

Replication of the basic models has been chosen as the preferred way to extend practical problems. This approach has the advantage that realistic and calibrated models develop, which, from an hydrological point of view, represent larger water systems. Obviously, these larger water systems are not connected.

To assess the sensitivity to the way in which water systems are extended by replication, the Luttenberg water system is also extended by groundwater flow elements between the subsystems of each next replica of the basic model.

### 7.5.3  Results

Table 7.18 presents the analysis results of the method of replication of the basic models of the case-studies. The table indicates the number of variables and constraints included in the optimization problems.

The size of optimization problems built can be determined from Table 6.2 (Page 153). Strictly following the underlying descriptions, yields several constraints which, in principle, are unnecessary in the formulation of the optimization problem. Examples are: inequality equations for horizontal line segments of damage functions, which yield $\eta \geq 0$ (Eq. 6.16); and continuity equations for weirs which cannot discharge because the water level is below the crest, which yield $Q_w = 0$ ($\kappa_w = 0$ in Eq. 6.44).

The unnecessary constraints are removed from the original optimization problem. This process is called 'filtering' here and is sometimes also indicated in literature as 'preprocessing'. Filtering of variables was considered impractical, since it changes the size of the optimal solution vector ($\overline{x}^*, \overline{u}^*$), which is supplied by the solver each run, and thus destroys the structure required to use the warm-start option. Filtering out unnecessary constraints reduces in the number of constraints by 25% to 40% in the models analyzed. The reduced number of constraints is indicated in Table 7.18.

The main evaluation parameters needed to determine the performance of interior-point solvers are the sparsity of the constraint matrix and the diagonal structure of the problem to solve (Terlaky, 1996). Despite the fact that solvers generally implement their own preprocessing methods, which affect the structure of the constraint matrix, here the original problem, as transferred to the solver, is analyzed. The way in which the optimization problems are built in the method developed is the same for all types of models, trying to obtain the smallest width in the diagonal of the constraint matrix. The width of the diagonal can increase in size by the number of variables in two time steps of the control horizon. This results from the continuity constraints for water quantity and water quality, in which both variables of the current and the previous time steps are included (e.g. Eq. 6.23).

The overall structure of the diagonals in the constraint matrix varies relatively little. Therefore, the sparsity of the original constraint matrix will be considered primarily in the following evaluations.

In optimization, sparsity of matrices is also indicated by the number of matrix coefficients unequal to zero, called 'nonzero elements'. The ratio of nonzero elements and the total number of elements was found to range from 0.3 to 1‰ in the models used.

The last columns of Table 7.18 indicate the time required for forward estimating only and the total time required for optimization including forward estimating. The results have been obtained by a regular present-day Pentium 133 MHz PC with 48 MB internal memory.

First of all, Table 7.18 shows that the sizes of the optimization problems, indicated by the number of variables, constraints and nonzero elements, increase nearly linear with the model

Table 7.18. Problem size and performance of optimization.

| Replication factor | Variables | Constraints [1] | Nonzero elements | Time forward estimating (s) | Total time optimization (s) |
|---|---|---|---|---|---|
| Delfland [2] ($T = 12$ steps; $\Delta t = 2$ hours) | | | | | |
| 1 | 948 | 920 | 2979 | 0 | 3 |
| 10 | 9480 | 9220 | 29606 | 4 | 51 |
| 50 | 47400 | 46250 | 148600 | 27 | 418 |
| 100 | 94800 | 92400 | 296900 | 63 | 1548 |
| Braakman ($T = 36$ steps; $\Delta t = 0.5$ hours) | | | | | |
| 1 | 972 | 837 | 2333 | 0 | 4 |
| 10 | 9720 | 8370 | 23330 | 5 | 48 |
| 40 | 38880 | 33480 | 93320 | 22 | 209 |
| 80 | 77760 | 66960 | 186640 | 79 | 817 |
| Luttenberg ($T = 12$ steps; $\Delta t = 4$ hours) | | | | | |
| 1 | 588 | 681 | 1980 | 0 | 2 |
| 10 | 5880 | 6810 | 19800 | 3 | 31 |
| 50 | 29400 | 34050 | 99000 | 14 | 185 |
| 100 | 58800 | 68100 | 198000 | 79 | 753 |

[1]   Unnecessary constraints filtered out
[2]   Including two water-quality variables

replication. The time required for forward estimating almost increases linear with the problem size for the models that were replicated 40 to 50 times. This results from the number of linearization actions required that increase linear by the number of side constraints. The total time needed for forward estimating is generally 5 to 10% of the total time required for optimization.

Figure 7.43 shows the three components that together make up the total time needed to solve the optimization problem for the basic and replicated Delfland models. The small contribution of forward estimating in the total time required to solve the Delfland optimization problems can be observed clearly. Another interesting phenomenon can be observed in the graph near 120,000 nonzero elements in the constraint matrix. From that value onwards, the time required for matrix updates increases considerably. This is caused by the limited amount of internal computer memory available. The memory required for large optimization problems imposes the greatest restrictions on the performance of the Constraint Manager. If required, the program extends its memory by creating and using virtual memory on disk, which obviously requires additional processing time. If a computer with 64 MB of internal memory is used, the kink in the curve was found to occur at a higher number of nonzero elements.

The curves in Fig. 7.44 show an overview of the results presented in Table 7.18 and others. The curves show that the increase in total optimization time first follows an almost linear function and is subsequently more similar to a power function of the problem size, expressed in the number of nonzero elements in the constraint matrix.

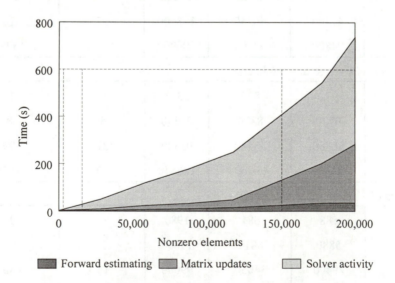

*Fig. 7.43. Components that together determine the time used for optimization: forward estimating, matrix updates and solver activity (Delfland basic model).*

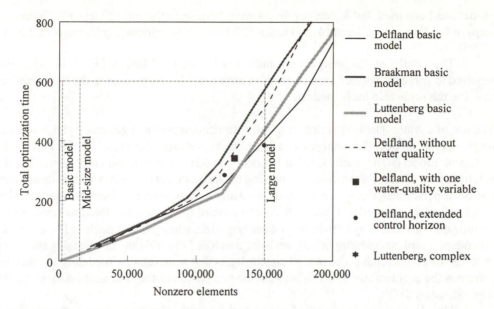

*Fig. 7.44. Total time used for optimization as a function of the optimization-problem size,*
*expressed in nonzero elements in the constraint matrix.*

Reading the problem sizes which can be solved at the Y-axis value of ten minutes (600 seconds), the number of nonzero elements proves to range from 150,000 to 180,000. This result corresponds to 750 to 1500 subsystems in optimization, 1000 to 1500 flow elements and 1000 to 1500 damage functions that together use 4000 to 6000 line segments.

In the remainder of this evaluation, the sizes of the models in optimization are also included for comparison. The following model sizes have been defined (Fig. 7.44):
• generalized (basic) model: approximately 10 subsystems and 15 flow elements;
• mid-size model: approximately 100 subsystems and 150 flow elements;
• large model: approximately 1000 subsystems and 1500 flow elements.

## 7.5.4 Sensitivity Analysis

The results obtained in the previous section are analyzed further by carrying out additional runs with models, which include the following characteristics:
• variation in the number of time steps in the control horizon;
• variation in the number of water-quality variables in the optimization model;
• increased model complexity.

The results of the sensitivity analyses are presented in Table 7.19. The first analysis is done to determine the sensitivity to the length of the control horizon. To do so, the basic Delfland model has been used, but instead of the 24-hour horizon including 12 steps of 2 hours, 48 steps of 0.5 hour were chosen ($\Delta t_b/4$) and a 120-hour horizon including 30 steps of 2 hours ($T_b \times 5$).

The results of the analyses are indicated by the two dots in Fig. 7.44. The time required to solve the optimization problem in both cases is comparable to the time required to solve the replicated basic models.

The use of a water-quality description in the optimization problem generally yields a rather large increase in both the number of variables and constraints (around 20% for both in the Delfland model). The basic Delfland model includes two water-quality variables. To determine the effect of water-quality modeling on the optimization time, a sensitivity analysis has been performed, excluding one and two water quality variables from the basic model. The results are presented in Table 7.19. As can be read from the table, the time required for optimization is reduced by 18% in water-quality modeling using one variable (418 versus 341 seconds at a replication factor of 50) and by as much as 52% without water-quality modeling (418 versus 200 seconds). The overall conclusion is that water-quality modeling in this case increases the problem size by 20%, whereas the time required for optimization increases by approximately 100%.

The dashed line in Fig. 7.44 presents the results obtained when water quality is excluded from the Delfland model. Surprisingly, the solver handles the problem including water quality more efficiently than without. This probably results from the relatively advantageous diagonal structure that develops as a result of the continuity constraints involved, which can be handled with relative ease by the solver. This is once more stressed in the graph by the rectangular point, which is located in-between the other curves of the Delfland model and shows the result including only one water-quality variable in the model.

As mentioned above, the replication method results in several models which are not mutually connected. To determine whether replication can be used to determine the performance of large models, the Luttenberg basic model has been extended by replication and subsequent linking of the separate models, each by nine extra groundwater-flow elements. This increases the complexity of the model and, moreover, presents a more realistic structure of the optimization problem than found by replicating basic models. The model with a replication factor of 16 (Table 7.19) represents a hilly area of approximately 60,000 ha with 208 interconnected subsystems and 331 flow elements.

Apart from increasing the complexity by this operation, the size of the problem increases as well. Table 7.19 and the stars in Fig. 7.44 present the results obtained. The stars in the figure show that the complexity of the problem slightly increases the optimization time required. The numerical results indicate that the optimization time required to solve the extended and more complex Luttenberg problem, is approximately 10% longer.

The analysis above demonstrates that a good impression of the total solution time of a complex model, consisting of hundreds of subsystems, can be obtained by first building a relatively small model with a representative ratio of subsystems and damage functions, and subsequently correct for size and complexity. This should be done with care though and preferably on the basis of the size of the optimization problem.

Despite the differences between the curves in Fig. 7.44 described above, the general conclusion of all curves together is that, especially for small to mid-size problems, the solution time for a new model can be estimated rather accurately from the graph.

A last sensitivity analysis has been carried out to determine the efficiency of the warm-start option of the solver used. It should be noted that, similar to the forward-estimation method (Sec. 6.2.2), the optimal solution of the previous cycle is shifted one time step to obtain the best starting point for the search in the current time step.

The results were found to depend largely on the size of the optimization time step and the accuracy of the predicted hydrological load and ranged from 0 to 50% reduction in solving time. These results can be explained by the fact that the warm-start option, in principle, requires that the solution of the previous optimization run is very close to the optimal solution for the current time step. If the optimal solution differs too much, the efforts of the search algorithm when starting its search from the previous optimum are similar to those when starting from scratch.

Table 7.19. Problem size and performance of optimization, sensitivity analysis.

| Replication factor | Variables | Constraints [1] | Nonzero elements | Time forward estimating (s) | Total time optimization (s) |
|---|---|---|---|---|---|
| Delfland ($T_b$ = 12 steps; $\Delta t_b$ = 2 hours): extending the number of time steps | | | | | |
| 10 / $\Delta t_b$/4 | 37920 | 37190 | 120756 | 17 | 286 |
| 10 / $T_b$×5 | 47400 | 46220 | 149454 | 21 | 380 |
| Delfland ($T$ = 12 steps; $\Delta t$ = 2 hours): excluding water quality variables | | | | | |
| 1 | 660 | 624 | 1920 | 0 | 2 |
| 10 | 6600 | 6260 | 19280 | 5 | 34 |
| 50 | 33000 | 31200 | 96000 | 19 | 200 |
| 100 | 66000 | 62800 | 193600 | 138 | 906 |
| Delfland ($T$ = 12 steps; $\Delta t$ = 2 hours): one water quality variable | | | | | |
| 50 | 39600 | 38050 | 128350 | 24 | 341 |
| Luttenberg ($T$ = 12 steps; $\Delta t$ = 4 hours): increased model complexity [2] | | | | | |
| 12 | 8244 | 9360 | 29172 | 3 | 53 |
| 16 | 11028 | 12516 | 39060 | 4 | 71 |

[1]   Unnecessary constraints filtered out
[2]   Replicated models mutually linked by 9 groundwater-flow elements

## 7.6    Concluding Remarks

This section summarizes the main conclusions of the three practical case studies and the efficiency of the simultaneous simulation and optimization method presented in the preceding sections of this chapter.

### Control Modes

Various control modes and combinations of control modes have been investigated in the case studies. Special comparisons were made of manual local control, local automatic control and dynamic control. The manual dynamic control mode is only suitable in practice if a limited number of regulating structures is included, because operators have to implement the control strategy determined by the Strategy Resolver of the DSS themselves. In all other cases, automatic dynamic control is the only possible option to use this advanced mode of control.

In comparison with manual control in the three practical case studies, local automatic control shows considerable improvements in water-system performance.

In comparison with local automatic control, dynamic control shows the greatest improvements in polder areas. Good results can also be achieved in hilly areas, however, water control by means of regulating structures should be possible. This is, obviously, a general phenomenon: to control water systems, sufficient water and regulating structures should be available.

With respect to water-quantity control, making better use of the current system capacities, generally means operating the regulating structures at the right moment and simultaneously preventing unnecessary operations. Short rain storms with a high intensity and long rainy periods with low intensity can be anticipated. No accurate prediction data are needed to determine control strategies to reduce considerably the chance of water-system failure.

Dynamic control can be forced to anticipate on precipitation events, even if no real-time predictions are available. In practice, this creates a surplus storage volume, prepared by an operator or control system, to avoid possible flooding events.

### Polder and Hilly Areas

On the basis of the results obtained, it can be concluded that polder areas can generally be controlled better than hilly areas, because the processes that primarily determine the success of control can be influenced better.

In hilly areas and especially the ones with a highly permeable subsurface, the main components of the water balance are: the outflow to adjacent water systems; infiltration into low-lying aquifers and shallow seepage throughout the water system. Shallow seepage can be controlled to some extent, however, the two other components cannot. For these reasons improvement in water-system control is more difficult to obtain in hilly areas in general. One of the options to increase the controllability of the hilly water systems is to increase the controllable components in the water balance, by increasing the number of regulating structures and by supplying alien water.

A complicating factor in hilly areas is that the slopes of the soil surfaces and water courses impose restrictions on the area which can be influenced by a regulating structure. For that reason relatively many regulating structures are needed to control the water system in both the lower and higher regions of hilly areas. Fortunately, these structures are relatively simple and generally need to have low capacities only.

### Control Horizon Required

The control horizon used in dynamic control is determined by the runoff characteristics of the water system considered. In general, water systems with rapid runoff characteristics, such as found in most polder areas, short control horizons of hours to days are sufficiently long. This has been shown by controlling the Delfland water system using a control horizon of 24 hours. In another case study of Fleverwaard, a control horizon of 12 hours proved to be sufficient to effectively control the polder area (Botterhuis, 1997).

Using the techniques developed, very slow-reacting water systems such as the Braakman water system (De Drie Ambachten) may also require relatively short control horizons. In those systems, the length of the control horizon is mainly determined by the capacity of the regulating structures that determine the controllability of the system. In the De Drie Ambachten case, the capacity of the Braakman Sluice, when it can be operated, is relatively large. However, if the capacities of regulating structures are smaller, the water system can be expected to require a longer control horizon.

Dynamic control of the Luttenberg water system has been analyzed using a control horizon of two days. This was found to be sufficiently long. Another case study has been performed for the hilly area of De Mark en Weerijs (Versteeg, 1997). This water system has a much less permeable subsurface. That case study proved that a control horizon of four days was required to obtain results in dynamic control that are comparable to the Luttenberg study.

### Water-Quality Control

The practical case study of Delfland has shown that dynamic control enables a combination of water-quantity and water-quality control. Water quality can only be influenced indirectly via regulating structures in such water systems, where dilution and transport are the main processes. Any water-quality control in these systems results from water-quantity control. This result can be generalized and is not only valid for the water-quality variables chloride concentration and BOD concentration, but also for other variables.

### Implementation of Dynamic Control

Dynamic control requires a central computer-based system. If the local automatic control systems are tuned well, it is not necessary to continuously control the regulating structures dynamically. In practice, the dynamic control mode may only be required for short periods.

The necessity of continuous dynamic control depends on whether conflicting requirements have to be satisfied. If the weighing mechanism included in dynamic control continuously has to balance the operations of regulating structures, the dynamic control mode should be used all the time.

The Strategy Resolver, which runs on the central computer, requires a precipitation prediction to determine whether pumping stations should be switched from local to dynamic control. As shown in the 30-year time-series calculations of the Delfland case study, optimal control of the water system could be obtained by switching to dynamic control only 6.6% of the time, using the dynamic-control threshold. This result was obtained for combined water-quantity and water-quality control. If the Water Board opts for water-quantity control only, dynamic control is needed only a fraction of this time. However, in the De Drie Ambachten and Salland cases, continuous dynamic control proved to be necessary, because the interests have to be balanced all the time to obtain optimal control strategies.

### Speed of Solving the Optimization Problem

The speed of solving the optimization problem is considered an important issue in this thesis. The main reason is that the speed required to determine operational control strategies, should be considerably faster than real time. The optimization problems that were built by the Constraint Manager of AQUARIUS on the basis of the generalized models of the case studies, could be solved within seconds. These results were obtained by a regular present-day 133 Mhz Pentium PC with 48 MB internal memory. If the generalized models are extended to large detailed ones, the solution time increases almost linear, when sufficient computer resources are available.

Implementation of the methodology in a real-time control system will require more detailed modeling. If, for example, the Delfland model is extended to include all polders and regulating structures of the water system, the solution time will increase to approximately one minute. This solution time definitely meets the requirements of real-time implementation, since the results of the Strategy Resolver are only required every hour or at even longer intervals. The results found for Delfland can be considered an upper limit for the De Drie Ambachten and Salland cases, since the detailed models of these systems are smaller.

Apart from the speed considerations in real-time control, speed of solving the optimization problem is also of importance for analysis purposes. The time needed for simultaneous simulation and optimization for the 30-year periods was several hours to days. This may limit the number of analyses which can be performed while automatic handling of runs and output should be considered necessary.

### Further Research

If, for any reason, dynamic control cannot come into action when it is needed, local automatic control should be available as back-up to guarantee safe operation of the water system. Possible reasons for dynamic control not coming into action are: communication failure between the central and local automatic control systems or a computer break-down. Such situations generally only occur for short periods and would only result in poor performance if they coincide exactly with an extreme system load.

Further research is required to assess to what extent the effect of failures of the control system would counteract the advantages of dynamic control. A preliminary assessment has been made for the Delfland reference polder, which is described in Sec. 8.3.2.

# 8 Water-System Design

## 8.1 Introduction

In Chapter 7, several examples were shown of the application of dynamic control in existing water systems. Dynamic control proves to be advantageous in operational water management and results in improved water-system performance.

On the basis of these results, the question arises whether dynamic control can contribute to avoid or postpone infrastructural investments. For sewer systems, the application of state-of-the-art control techniques in several European countries proved to reduce or avoid large investments (e.g.: in Spain: Cabot Plé, 1990; in Germany: Khelil & Schneider, 1991; in Sweden: Schilling, 1996; in Denmark: Babovic, 1991 in the Netherlands: Bakker et al., 1984). However, until now, the starting point has been to enhance the operation of existing systems, not to redesign water systems to include the application of advanced control methods.

The limitations of standard regional water-system design have been reported by several authors. A clear inventory of problems inherent to this kind of design and a listing of several improvements needed are given by Bouwmans (1994).

The principle of some standard methods will be described in the first part of this chapter. The second part of the chapter aims to show a modern alternative to the standard design methods.

## 8.2 Methods Applied

Capacities of water systems are designed with the help of general guidelines with respect to discharge, surface-water storage and frequencies of extreme water-system loads. Several standards have been set for the design of water systems including rural areas and combined rural and urban areas (e.g. Grotentraast, 1992). The responsible water authorities generally interpret these standards and apply them according to the specific circumstances in their water system.

In the following, some design methods will be discussed that are used for the determination of water-system capacities in rural and urban areas. In practice water-system dimensions are determined by:

- the capacity of discharge structures;
- the storage capacity of the surface-water subsystem.

If the available storage is small, the discharge capacity has to be large. If the storage is large, only smaller discharge capacities are needed. This principle is shown in Fig. 1.6 (Page 8).

Two main methods are applied in water-system design at the moment:
1.   design on the basis of critical discharge;
2.   design on the basis of dynamic discharge.

The first method incorporates empirical standards and is used widely, since it is simple and it does not require many location-specific data. The second method incorporates a form of water-system dynamics, which does require location-specific data and also data on the water-system load in time.

## 8.2.1  Critical Discharge Method

This section describes a discharge and capacity design method traditionally applied in the Netherlands. A detailed comparison between the various methods available can be found in Schultz (1992). The method described uses empirical standards for pervious surfaces and precipitation duration curves for impervious surfaces. Two variants of the method are discussed here: a general method, used to design rural water systems only, and a method used for combined rural and urban water systems.

### *Pervious Surfaces*

The method applied to design water systems in rural areas with pervious surfaces, uses the general critical discharge, which only depends on soil texture and land use. Critical discharge values have been published by several authors (e.g. Van der Molen, 1992; Grotentraast, 1992). The method is applied especially when few or no data on discharge are available. On the basis of soil type, land use and groundwater-level development a critical discharge is selected, e.g. 10 mm/d.

In this case, the critical discharge is considered to be the discharge which is exceeded no more than once a year. Regulating structures are designed on the basis of this discharge. In case of high system loads, where the regulating structures cannot discharge all excess precipitation immediately, the surface-water level rises.

### *Combined Pervious and Impervious Surfaces*

This method incorporates runoff from rural and urban areas, that comprise both pervious and impervious surfaces, and determines the discharge capacity and storage for the combined water system.

If the storage is known, the discharge capacities of structures can be determined, or if the discharge capacities of structures are known, the storage can be determined.

On the basis of standards, a critical discharge is selected for the pervious surface, which is exceeded with a specific frequency, e.g. once every 10 years. For the impervious surface, a precipitation duration curve is selected, e.g. a curve which is considered to be exceeded less frequently than once every 10 years. The assumption is that precipitation runs off from impervious surfaces instantly and no infiltration occurs. The following example demonstrates how this method is applied.

Figure 8.1 graphically shows how surface-water storage can be determined for a water system that comprises both pervious and impervious surfaces. In the example, the discharge capacity (e.g. of the draining pumping station) is given and represented by a line in the graph (e.g. 8 mm/day).

For the pervious surface, a critical discharge, is selected from standards. That discharge is considered to be exceeded with a specific frequency (e.g. 20 mm/day once every 10 years). This value may also incorporate seepage from deep groundwater and discharges from other areas.

For the impervious surface, a precipitation-duration curve is selected which represents discharge to surface water, which is exceeded with the same frequency (e.g. also once every 10 years).

The discharges from the pervious and impervious surfaces are added, together called the *system-load curve* (Fig. 8.1).

Finally, the moment when the difference between the pump-discharge line and the system-load curve is largest, is determined. That moment is called the 'critical-discharge time' and represents the storage quantity required in the system.

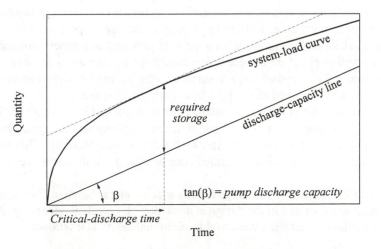

*Fig. 8.1. Design example for water-system capacities, assuming combined discharge from pervious and impervious surfaces.*

The critical discharge time shows how long the water level will rise during and after precipitation under design conditions. In the design, this period is generally kept short and within practical limits (e.g. one day), to prevent long periods of extremely high water levels.

The water storage required, is distributed between surface-water and groundwater subsystems, but it is common practice to consider only the surface-water subsystem. If the maximum water-level rise allowed for the surface-water subsystem is known, the area needed for the surface-water subsystem can finally be determined.

As in the example, the frequency of exceeding the maximum water level is generally considered equal to the frequency of exceeding the selected rural discharge and precipitation duration curve.

The design method described, incorporates several assumptions that are theoretically incorrect and, in general, accumulate various safety margins, of which the main ones are listed below.

- The discharge curve for the impervious surface is considered to equal the precipitation duration curve. This is incorrect. The duration curve is an envelope curve and its shape does not represent a real precipitation curve. The discharge curve of the impervious surface is generally lower as a result of interception of precipitation, infiltration in semi-impervious surfaces and inertia processes in runoff (Van de Ven, 1983).
- Critical precipitation is assumed to fall at the same moment on pervious and impervious surfaces. This is practically impossible, since the critical situation for the rural pervious area generally occurs in winter during long consecutive periods of precipitation with a relatively low intensity, while the critical situation for the impervious urban area generally occurs in summer, during a short period of precipitation with a high intensity.
- Continuous runoff from the pervious surface is assumed. In reality, this only occurs during long precipitation events with a relatively low intensity. Delay effects resulting from water flows from the various surfaces to the surface-water subsystem are not incorporated. In practice, a rather large delay effect can be present.
  Unequal critical discharge times for areas with pervious and impervious surfaces can have a positive effect on the combined runoff from these areas in time. Figure 8.2 demonstrates this: the discharge to surface water (in mm/h) of the combined area is lower than the combined discharges from both separate areas.
- The initial situation in the water system is not incorporated in the method. Therefore, the implicit assumption is that the soil is at field capacity and the groundwater level is continuously high in comparison to the surface-water level. This assumption neglects the storage in the unsaturated zone and the groundwater system.

Because of the effects listed, the discharge to the surface-water subsystem is exaggerated and the safety margin incorporated in the design is most likely larger than necessary. A problem of this design method is that the extent of this additional safety is unknown.

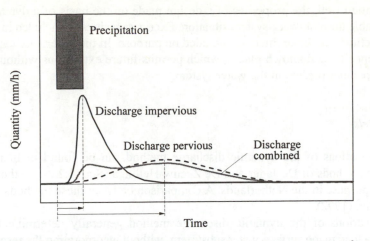

*Fig. 8.2. Combined discharge from pervious and impervious surfaces.*

## 8.2.2  Dynamic Discharge Method

Designs based on the dynamic discharge method, require a dynamic model that represents the actual or future situation in the water system. A sufficiently detailed model is needed to determine the runoff quantities as a function of time. In general, such a model is run on the basis of time-series data on hydrological-load variables such as precipitation. The runoff model describes the flow processes in the soil, groundwater and surface water in more or less detail. A detailed model, for instance, includes unsaturated-zone flow, whereas a less detailed model neglects these processes and assumes a continuous situation of field capacity.

Discharge measurements and, in principle, also groundwater-level measurements are needed to calibrate the runoff model. Preferably, field discharge measurements are used for this purpose, since this enables determination of specific water-system parameters. If such data are not available, discharge measurements of regulating structures can also be used.

In general, computer-based tools are a prerequisite for efficient dynamic discharge modeling and analysis of the water system. If a calibrated model of a water system is available, extremes analysis is possible, using time-series data or only a set of extreme events. In this way, a more detailed picture of the water system's reaction to loads can be obtained.

Both deterministic and black-box modeling techniques can be used to determine the system-load-response function. From this function, a critical discharge can be derived which includes location-specific data.

The dynamic-discharge approach circumvents some of the major disadvantages of the general critical discharge method described in Sec. 8.2.1.

If applied well, the properties of a design made on the basis of a dynamic model, closely resemble the real water-system situation. Excess capacities as included in the critical discharge method, can be omitted, or included on purpose. In the latter case capacities are included at specific and known places, which permits future extensions without having to adjust the capacities present in the water system.

### Runoff Models

Various applications of the dynamic discharge method can be found in literature. The deterministic methods of De Jager and De Zeeuw-Hellinga (Sec. 4.2.2) are the main ones used for this purpose in the Netherlands. A comparison between these methods is given by Van der Molen (1992).

Applications of the dynamic discharge method generally determine the system discharge as a load to the surface-water subsystem, without incorporating the response of the surface water on the groundwater subsystem. If, for instance, as a result of a high discharge, the surface-water level rises, the effect of possibly restricted soil outflow is not taken into account. For this reason, these methods are generally only applicable for rural-area modeling and to describe the system behavior during the winter season, when outflow from the soil prevails.

The current water systems comprise both rural and urban areas. Moreover, critical loads on urban areas occur in summer. This is an important shortcoming of these applications of the dynamic discharge method.

### Integrated Models

Integrated dynamic discharge models include the various subsystems and the regulating structures of a water system. The use of integrated models has the advantage that the runoff from all types of surfaces can be included. Furthermore, the interaction between groundwater and surface water can be taken into account in detail.

Designing on the basis of a dynamic model of the water system, prevents unnecessary excess capacities, which were automatically included in the designs according to the critical discharge method.

The principle of these applications is to simulate long time series, e.g. 30 or 50 years and to count the number of undesirable water-system situations, called failures. Such a design should be assessed with care, however.

The design assumes a continuously operational water system. If the water system is configured exactly as modeled, if the structures are never out of order and if the real system parameters are exactly as assumed, the frequency of failure of the water system will be the same as in the design. This is generally not the case, however. They present a far too optimistic result in terms of water-system capacities and therefore the stochastics of the availability of control-system elements should be included specifically.

## 8.3    Dynamic Design Procedure

This section presents a variant to the dynamic discharge method as an alternative to the critical discharge methods. Since they are widely applied, the critical discharge methods will here be called 'standard design methods'.

The dynamics of the water system, the interests present and the hydrological load on the water system are here included in the design and for that reason the method developed is called a *dynamic design*. An important difference between standard design and dynamic design is that the latter includes an integrated model of the water system and location-specific data. Furthermore, the way in which the water system is controlled is explicitly incorporated.

The starting point for regional water-system design using standard methods, is the critical failure frequency of the water-system. Events such as flooding and overflowing are considered failures. A dynamic design, however, is evaluated by determining the failure frequency. In this respect a *system failure* is defined as a situation in which the requirements of one or more interests are violated in such a way that the objectives cannot be met, e.g.: flooding, fish mortality, loss of harvest.

The dynamic design procedure is introduced step-wise. The performance of a specific water system and its control system are evaluated to determine whether the system satisfies the requirements of the interests involved.

Given the capacity limitations of all water systems, each water system fails to satisfy the requirements of a particular interest with a certain frequency. In the dynamic design procedure, for each interest a specific 'critical failure frequency' can be included, i.e. the failure frequency which should not be exceeded for that interest. The dynamic design procedure can additionally incorporate other evaluation parameters such as the duration of failures and the performance index of the water-system.

It should be noted that the procedure presented here, differs fundamentally from the risk-based approach which is traditionally used in the design of river and sea dikes in the Netherlands. That approach, focuses mainly on the balance between the probability of a failure occurring and the associated damage in terms of costs.

The design approach presented in this thesis, defines interests, without explicitly assigning costs to failures. Similar to dynamic control, this permits incorporation of subjective components. Interests of which the value is difficult to define in terms of money, can thus be incorporated with relative ease.

### 8.3.1  Step-Wise Procedure

An exact recipe for the design of regional water systems, including system dynamics, cannot yet be given. This requires comparison of the results of the present study, with practical data. The latter, however, could not be found in literature, while collecting them in practice was outside the scope of the present research. At the moment, advantages and disadvantages can therefore only be based on thought experiments and computer calculations. This section should therefore be seen as an attempt in structuring the elements of the dynamic design procedure.

The following steps are distinguished in the dynamic design procedure:
1.    standard design,
2.    water-system modeling,
3.    identifying failure mechanisms,
4.    time-series calculations,
5.    evaluation,
6.    sensitivity analysis.

The flow-chart in Fig. 8.3 shows all steps of the procedure. In addition, a general description of each step is given below. In principle, the procedure can be used to design a water system on the basis of any interest of which the requirements can be expressed in water-system variables (Sec. 3.1). A practical example is given in Sec. 8.3.2.

## 1. Standard Design

If a design has to be made from scratch, the first step in the procedure is to create a design, using one of the standard methods (Sec. 8.2). This is necessary since the dynamic design procedure is a verification method in which an existing design is adjusted iteratively to meet all requirements as set. To start with, the basic outline of the design should be determined, such as the general layout and developments planned in and outside the water system.

The critical discharge method is used to create the initial design, called the 'standard design', which uses standards for the critical failure frequency. If an existing water system has to be redesigned, the existing layout can be used as a basis. This situation arises when the arrangement is to be changed, the weights of existing interests are reconsidered or other interests are now considered important.

The dynamic design requires to set the critical failure frequency for each water system interest $E_{fc}$, e.g.: the frequency of flooding should be below 0.1 a year, the frequency of violating ecological interests in terms of water pollution should be less than once a year. These frequencies are used as evaluation parameters in the evaluation step (5).

The critical frequency of flooding is generally the only frequency used in the standard design. To enable comparison of the results, the critical frequency of the common-good interest of flood-prevention should be the same in the dynamic design as in the standard design.

## 2. Water-System Modeling

The second step of the procedure consists of building a model of the water system, the general dimensions of which are based on the standard design. This model should be an accurate representation of reality. Therefore, calibration of the model is required. Alternatively, calibration results of similar water systems can be used. General remarks on this subject are given in Sec. 7.1.4.

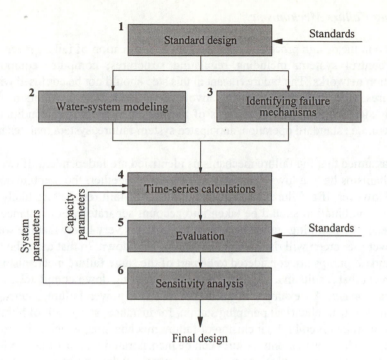

*Fig. 8.3. Steps of the dynamic design procedure.*

Elements of the water system that determine its failure should be modeled carefully, since in the following steps, such failures play an important role. Therefore, this step and the parallel step of identifying failure mechanisms (3), are closely related.

In the next step (time-series calculations), various control modes for regulating structures will be used. Combinations of control modes should be defined in sets, ranking from simple to advanced. These combinations should be included in the model in such a way that sets that use advanced modes, which depend on fail-critical devices, can be replaced by more simple ones, which do not depend on these devices. An example of two control-mode combinations could be: mixed local control and dynamic control by a central computer (advanced set); local control of all regulating structures (simple set). The set of control modes which is normally used is called the 'regular set' of control modes.

In the modeling step, a parameter analysis should be performed. On the basis of the accuracy of the available model data and the known accuracy of developments in and outside the water system, the parameters that are expected to affect the design most should be identified. These parameters are used in the sensitivity analysis step (6).

## 3. Identifying Failure Mechanisms

The third step in the design procedure comprises the identification of failure mechanisms in the entire control system, including regulating structures, computer equipment and communication networks. The failures meant in this step should not be confused with water-system failures, since failure mechanisms involve only devices or operators of the water system. This step requires careful analysis of the possible operational conditions of the control system, i.e.: standard operation, anticipated system failure, system maintenance (Sec. 2.4.1).

It is assumed that the failure mechanisms identified are independent. If two or more failure mechanisms have a low probability of occurring together, the combination can be neglected. However, if a failure mechanism is built up of failures that are likely to occur together, this combination should be taken into account separately. For instance, when a garbage screen of a pumping station becomes clogged, it is very likely that the water-level difference over the screen will rise so high that pumps shut down. In that case, the failure of the screen and all pumps are considered to be part of the same failure mechanism.

An event that results in a regulating structure breaking down completely, is called a *common cause event*. An example of such an event is a power failure, interrupting the electricity supply to an electrical pumping station, for instance, as a result of lighting.

A problem involved in the inclusion of failure mechanisms, which is inherent to the method presented here, is that only failures can be incorporated that last for a period which is of the same order of magnitude as an entire water-system-failure event or longer, e.g. hours to days. A mechanism which results in failures that last only a few minutes is neglected.

The availability of regulating structures or units of these structures related to failure mechanisms, is represented by 'failure scenarios' in the time-series calculations step (4), e.g.: no units out of order; one unit out of order, etc.

## 4. Time-Series Calculations

The fourth step in the procedure consists of time-series calculations that describe the reaction of the water-system to hydrological loads. Hydrological time-series data over long periods, e.g. 30 years, are required. If such data are not available, data from other locations can be used. However, it should first be verified whether these data are representative for the area for which the new design is intended.

The first series of calculations is for the simplest set of control modes (e.g.: local manual control or local automatic control), followed by advanced sets as defined in the modeling step. Most likely, these time-series calculations prove that the standard design contains unused capacities. However, this is not necessarily the case.

It is assumed that the average frequency of a failure can be determined from the number of failures recorded during a time-series calculation. This frequency is called the 'recorded failure frequency'. If, for instance, a system fails three times within the analyzed period of 30 years, the recorded failure frequency is 0.1 a year.

In addition to time-series calculations assuming all regulating structures fully operational, calculations are required that assume possible combinations of failure scenarios and sets of control modes. An example is a time-series calculation in which a pumping station works at a reduced capacity as a result of one pump being out of order, combined with an advanced mode of control.


## 5. Evaluation

In the evaluation step, results of the various time-series calculations are used to determine the expected failure frequency of all interests. Several assumptions are made:
- during a failure of the control system, the regular set of control modes (advanced) can be replaced by a more simple set of control modes;
- before each system load, which leads to a failure, the water system is in an equilibrium situation, i.e. satisfying all interest requirements;
- failures result only from water-system loads;
- the failure scenarios distinguished occur independently.

Using these assumptions, the expected failure frequency of the water-system $E_f$, can be expressed by:

$$E_f = \sum_i^m P(CM_i) \sum_j^n P(FS_j | CM_i) \cdot S_{f_{i,j}}, \tag{8.1}$$

in which:
$E_f$     : expected failure frequency the water-system ($y^{-1}$);
$i$     : counter for control-mode combinations;
$m$     : total of control-mode combinations;
$P$     : probability;
$CM$     : control-mode set;
$j$     : counter for failure scenario;
$n$     : total number of failure scenarios;
$FS$     : failure scenario;
$S_f$     : recorded failure frequency ($y^{-1}$).

On the basis of experience with the current water system or similar water systems, the probability of possible failure scenarios can be determined. If the required data are not available, estimates have to be made on the basis of their expected frequency and duration. The sensitivity of the design to these assumptions has to be included in the sensitivity-analysis step (6).

On the basis of time-series calculations assuming one or more failure mechanisms, $E_f$ can be determined and compared with the critical frequency of system failure $E_{fc}$. If these do not match, another iteration is required to obtain better capacity parameters, e.g. a smaller pumping capacity; a larger inlet capacity; a smaller surface-water subsystem.

The duration of a failure can greatly affect the damage to interests. Therefore, an extra criterion with respect to duration can be included in the analysis step.

A simple set of control modes can be used as a basis to determine the accepted duration of a water-system failure. An extra criterion applied for the advanced set of control modes could, for instance, be that the total duration of water-system failure should not increase when advanced modes are applied as regular control modes.

To determine failure duration from the time-series calculations, the following equation is used:

$$E_d = \sum_i^m P(CM_i) \sum_j^n P(FS_j|CM_i) . S_{d_{i,j}},$$ (8.2)

in which:

$E_d$     : expected failure duration of the water system (h/y);
$S_d$     : recorded failure duration (h/y).

With respect to the performance of the water-system an additional restriction could be that a design incorporating advanced control modes should not have a lower performance index (Eq. 7.1) than that using a simple control mode.

Summarizing, possible evaluation criteria for a design can be:
• the expected frequency should match the critical failure frequency;
• the regular use of an advanced set of control modes should not increase the duration of water-system failures;
• the regular use of an advanced set of control modes should not decrease the performance index of the water system.

Once these, or similar conditions are fulfilled, the dynamic design procedure can proceed to the next step. If not, the capacity parameters in the water-system model have to be adjusted and another iteration is needed.


## 6. Sensitivity Analysis

A sensitivity analysis should be performed for those parameters that the system proved to be sensitive to in the modeling step (2) and for data that were estimated.

In general, the parameters used in the regular set of control modes and the failure mechanisms with the highest probability of occurring and which have the highest impact, as identified in step (3), should be selected for the analysis.

The sensitivity analysis requires at least one extra iteration in the design procedure. Adjustments to the system parameters should be done with care and only if a combination of possibly incorrect assumptions seems logical.

Once all steps have been completed successfully, the design can be finalized and implemented if approved.

### 8.3.2 Dynamic Design Example

This section presents an example case which shows the possibilities and limitations of the dynamic design procedure, described in general terms in the preceding section. For demonstration purposes, the Delfland reference polder, described in Sec. 7.2.4 is used (Fig 8.4). All time-series calculations have been executed using AQUARIUS.

The reference polder is a typical glasshouse polder, with rapid runoff characteristics. The polder is drained by one pumping station. During periods of drought, water is let into the polder via several unregistered inlets and a controlled inlet structure.

As in the practical design of this polder (Delfland, 1991), only the flood-prevention interest is distinguished, which is related to the surface-water subsystem. This choice permits comparison between the standard design method and the dynamic design procedure.

Figure 8.5 shows a simplified interest-weighing chart, defined for reference polder. The figure is included to show that using the same method as described here, also permits other interests and water-system variables to be included, using the weighing methods developed for dynamic control.

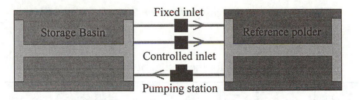

*Fig. 8.4. Schematic representation of the reference polder.*

The example design case is described on the basis of two design questions:
- whether the current drainage system of the polder meets the critical failure frequency of 0.1 a year and if not, what adjustments are required to the water system to meet that frequency;
- whether the total glasshouse surface area can be extended in the future and, if so, to what extent, and whether additional drainage capacity would then be required.

### 1. Standard Design

The first step in the procedure is to create a standard design or verify the available design, using the critical discharge method. Table 8.1 lists the data needed. A distinction is made between the summer and the winter season with respect to the critical discharge and precipitation duration curves.

If these data are applied in the critical discharge method for combined permeable and impermeable surfaces (Sec. 8.2.1), the critical water-level rise turns out to be 0.45 m for the summer season and 0.30 m for the winter season (Fig. 8.6). The design proves to meet the applicable standards exactly, since a rise in water level of 0.45 m brings the system on the brink of flooding.

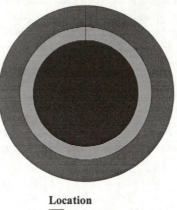

**Interest**
■ Flood prevention
**Key variable**
▨ Surface-water level

**Location**
■ Reference polder

*Fig. 8.5. Interest-weighing chart for the reference polder.*

Table 8.1. Basic data for the design of the reference polder.

| Subject | Properties |
|---|---|
| Area | 30 ha rural, 5 ha urban, 60.2 ha glasshouse, 4.8 ha surface water (total 100 ha) |
| Target water level Soil-surface level | reference level - 1.45 m reference level - 1.00 m |
| Pumping-station capacity | two pumps with total capacity of 0.2 m³/s |
| Inlet capacity | 0.04 m³/s unregistered (fixed open), 0.04 m³/s controlled |
| Interest considered Water-system failure regarding the interest | flood prevention surface-water level above soil-surface level |
| Critical failure frequency | 0.1 a year |
| Rural area design | critical discharge:    10 mm/day (summer) 20 mm/day (winter) |
| Urban and glasshouse area design | precipitation duration curves for summer and winter seasons, with a frequency of exceeding of 0.1 a year |

In the standard design, unregistered inlets are not included. Inclusion of these inlets would enlarge the required capacities. On the other hand, rainwater basins present in the glasshouse areas are not taken into account either. If these basins are included in the design, smaller capacities are required. The extent of these capacities does not follow from the standard design, however.

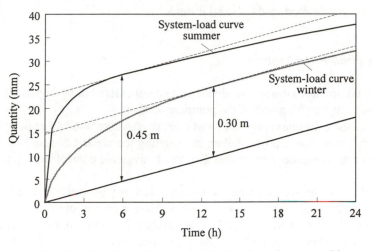

*Fig. 8.6. Standard design example for the reference polder.*

### 2. Water-System Modeling

The data of Table 8.1 were used in modeling together with additional runoff data from the rural and urban areas. These data were retrieved from calibrations of typical rural and glasshouse polders in the Delfland Area.

Corresponding to the actual situation in the Delfland Area, the unregistered inlets (in the model together called 'fixed inlet') are kept open continuously, while the controlled inlet is opened only when required. The fixed inlet considerably reduces the effective drainage capacity of the polder to 0.16 m³/s, a reduction by 20%.

The influence of rainwater basins in the glasshouse area is included in the model, assuming a storage capacity of 500 m³/ha for 50% of the glasshouses. The rainwater basins reduce the peak discharge onto surface water considerably, especially in summer when these basins are not entirely filled or even empty most of the time.

The following sets of control modes are defined:
1.    local automatic control of the pumping station and local automatic control of the inlet on the basis of the average water level in the polder;

2.      dynamic automatic control of the pumping station and local automatic control of the
        inlet on the basis of the flood-prevention interest, with the preferred water level equal
        to the target polder water level.

In this particular example, each set of control modes covers only one type of control: local
automatic or dynamic automatic. Therefore, the term 'set of control modes' is abstracted to
'mode', in the remainder of this section, and the term automatic is omitted. Thus the
following terms are used: 'local control mode', 'dynamic control mode'. The term 'regular
set of control modes' is shortened to 'regular mode'.

### 3. Identifying Failure Mechanisms

The following failure mechanisms are identified and considered to be important:
*   failure of one pumping unit of the pumping station;
*   failure of the entire pumping station as a result of a common cause: maintenance of
    the pumping station; a power failure interrupting the electricity supply;
*   failure of the computer system that determines dynamic control strategies.

Because only the flood-prevention interest is considered, any potential failure mechanisms
of the inlets are not taken into account. Their influence on the options that satisfy the
requirements of this interest are negligible in the design. However, if water-quality-related
interests, such as glasshouse horticulture would have been included in the design, the failure
mechanisms of the inlets should have been included as well.

### 4. Time-Series Calculations

Calculations for local and dynamic control of the water system were carried out, applying the
following scenarios:
a.      no pumps out of order;
b.      one pump out of order;
c.      all pumps out of order.

The time-series calculations were carried out in such a way that after each water-system
failure, the equilibrium water-system state was restored again. This ensures realistic water-
system conditions preceding system loads that lead to failures.
        The results of the calculations, using the De Bilt time series, are presented in Table
8.2. The table shows that using local control according to scenario (a) the total number of
water-system failures is only one in 30 years. Using dynamic control, zero failures would
occur in these 30 years.
        The recorded number of failures, if one pumping unit is out of order (b), is
considerably higher: 15 every 30 years using local control. This number decreases to 10 when
dynamic control is applied.

Water-system failures occur 267 times if all pumping units are out of order (c). This is irrespective of the control mode of the pumping station, obviously.

Table 8.2. Water-system failures in a 30-year time-series calculation.

| Scenario | Recorded number of water-system failures | |
|---|---|---|
| | Local control | Dynamic control |
| a. No pumps out of order | 1 | 0 |
| b. One pump out of order | 15 | 10 |
| c. All pumps out of order | 267 | 267 |

The probability of failures with no pumps out of order cannot be determined accurately on the basis of the data presented in the table (one and zero failures). To determine these data more accurately, the calculation results are analyzed further to determine whether the number of times a water level is exceeded, follows a logical sequence.

Figure 8.7 presents the results of one of the various time-series calculations which have been carried out to establish the required pumping capacity for the pumping station, explained in the evaluation step. The vertical axis represents the recorded number of times a specific water-level is exceeded. The straight line represents the linear regression result of that number, obtained by the least-squares method. Note that around the situation of system failure, only few data are available. Therefore, the recorded number of times the soil surface level was exceeded, may differ from the regression result (level of failure: ref-1.00 m).

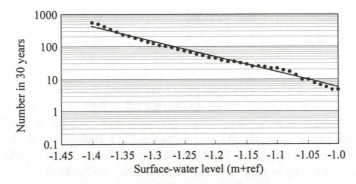

Fig. 8.7. Water-level exceeding in a 30-year time-series calculation for the reference polder (local control; no pumping units out of order; 65% of the original pumping capacity).

To answer the first design question, formulated at the beginning of this section (i.e. whether the current drainage system meets the critical failure frequency of 0.1 a year) a range of pumping capacities has to be incorporated in the model for each control mode. The results of the corresponding time-series calculations are presented in Fig. 8.8. The figure shows a gradual decrease in the number of system failures against an increase in pumping capacity, for four control mode-scenario combinations.

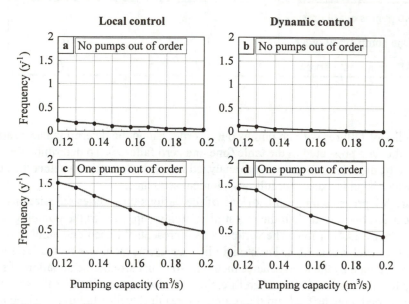

*Fig. 8.8. Recorded frequency of water-system failure as a function of pumping capacity, using local control (a and c) and dynamic control (b and d) (control-system failures not included).*

## 5. Evaluation

To determine the expected frequency of water-system failure, it is assumed that when the control system fails, the regular control mode (advanced) is replaced by a less advanced mode. In the current example, dynamic automatic control is active whenever possible. When the central computer breaks down, the dynamic control mode is replaced by the local control mode. When a local unit breaks down, local manual control is applied. The effect of local manual control is assumed to be similar to that of local automatic control.

The following average durations of each control-system failure are assumed:
- one pump out of order: one day a year;
- all pumps out of order (common cause): half a day a year;
- computer-system break-down (no dynamic control): five days a year.

The right-hand graphs (b and d) in Fig. 8.9 show the results of the evaluation using regular dynamic control. The results shown have been obtained by applying Eq. 8.1 and Eq. 8.2. To give an impression of the result obtained when using regular local control, the left-hand graphs present the results of local control.

From graph (a) in Fig. 8.9, an expected failure frequency of 0.05 a year can be read (the dotted line starting at 0.2 m³/s), using regular local control. When dynamic control is used (b), the expected failure frequency is only 0.02 a year.

These data provide an answer to the first design question: the current drainage system of the reference polder meets the critical failure frequency of 0.1 a year.

Now the question arises which should be the right capacity for a newly designed pumping station. To answer this question, time-series calculations have been executed, using equally reduced pumping capacities for the two pumping units of the pumping station. These results are also presented graphically in Fig. 8.9.

The graphs show that the capacity of a pumping station, to exactly meet the critical failure frequency, would have to be 0.172 m³/s in case of regular local control and 0.136 m³/s in case of regular dynamic control. Consequently, the conclusion is warranted that the system, if controlled locally, needs 14% less capacity than follows from the standard design. If dynamic control is used, including the possible failure mechanisms in the design, the standard design capacity could be reduced by 32%, provided that only the criterion of critical failure frequency applies.

The lower graphs (c) and (d) in Fig. 8.9 show the expected duration of failure. These data can be used to evaluate the consequences of the regular control mode to the duration of system

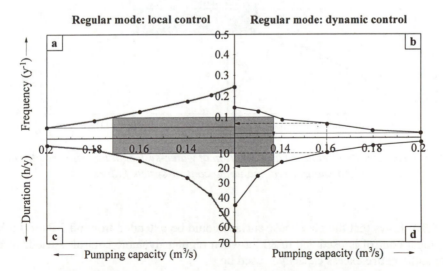

*Fig. 8.9. Expected frequency and duration of water-system failure as a function of pumping capacity, including control-system failure.*

failure. As can be read from the shaded part in the graph, the expected duration of failure, using regular dynamic control at a capacity of 0.136 m³/s, is 19 hours a year. If regular local control is used, the expected failure duration is 10 hours a year.

When, in addition to the frequency criterion, the expected duration criterion is applied, the pumping capacity for dynamic control becomes 0.160 m³/s, which means a reduction by 20% (instead of 32%) compared to the standard design. However, the failure frequency would be reduced to 0.07 a year (instead of 0.1).

The standard design was found to include excess capacity in all control modes. To answer the second design question (i.e. whether the total glasshouse area can be extended in the future) the excess capacities found, permit extension of the glasshouse surface.

The curves in the graphs in Fig. 8.10 show what happens when the present surface of 60.2 ha is gradually increased to 90.2 ha. The latter means that the entire rural area is covered by glasshouses. The curves show an increase in the number of water-system failures as a result of extending the glasshouse surface, if the current storage capacity of the surface water and the discharge capacity of the pumping station remain the same as their standard design values.

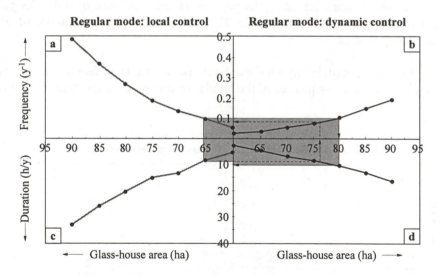

Fig. 8.10. Expected frequency and duration of water-system failure as a function of glasshouse area, including control-system failure.

Figure 8.10 shows that the glasshouse surface could be extended from 60.2 ha to 66 ha if regular local control is used and to 80 ha when regular dynamic control is used. In both situations, the entire excess capacity is used up.

The second design question can now be answered: no additional drainage capacity is required for the reference polder if the area currently covered by glasshouses is extended, provided this growth will be less than 10% if regular local control is applied and less than 33% if regular dynamic control is applied.

An extension of the glasshouse surface to 66 ha would result in an expected failure duration of 9 hour a year, if the regular control mode is local control. If the glasshouse area is extended to 80 ha and dynamic control is the regular control mode, failure is expected to occur during 10 hours a year. On the basis of these results it can be concluded that the duration criterion (again) imposes an extra restriction, if dynamic control is used, to the permissible extension of the area covered by glasshouses. The glasshouse surface could be extended to 77 ha (28% extension), the associated failure frequency being 0.08 a year.

Table 8.3 summarizes the results. In addition to the criteria of failure frequency and failure duration, the performance index is included. The table shows that in all cases, the performance index using dynamic control is higher than that using local control, which proves that in this particular case the performance index does not further restrict the pumping capacity required, nor the permissible extension of the area covered by glasshouses.

Table 8.3. Results of a dynamic design for the reference polder.

| Regular control mode | Pumping capacity $(m^3/s)$ / reduction [1] | Glasshouse surface (ha) / extension [1] | Performance index (-) | Failure frequency $(y^{-1})$ | Failure duration (h/y) |
|---|---|---|---|---|---|
| Local | 0.200 / 0% | 60.2 / 0% | 1 | 0.05 | 5 |
| Local | 0.172 / 14% | | 0.9 | 0.10 | 10 |
| Dynamic [2] | 0.160 / 20% | | 2.3 | 0.07 | 10 |
| Dynamic | 0.136 / 32% | | 1.7 | 0.10 | 19 |
| Local | | 66 / 10% | 0.8 | 0.10 | 9 |
| Dynamic [2] | | 77 / 28% | 1.9 | 0.08 | 9 |
| Dynamic | | 80 / 33% | 1.8 | 0.10 | 10 |

[1] Relative to standard design
[2] Using the failure-duration criterion

## 6. Sensitivity Analysis

In general, an analysis of various water-system parameters is needed to determine the sensitivity of the expected failure frequency to the accuracy of these parameters.

For the example case of the reference polder, the sensitivity to two parameters of the designed water-system properties was determined: the size of the rainwater basins, modeled in step two, and the duration of control-system failures, identified in step three. An analysis was performed to determine the sensitivity of the total extension of the glasshouse surface in the reference polder to these parameters.

First, the sensitivity of the design to the size of rainwater basins is analyzed. This includes the accuracy in establishing the total size of the rainwater basins from maps and field observations and the actual availability of these basins. In the analysis it is assumed that the size of 500 m³/ha represents an overestimate by 50 m³/ha.

Second, the sensitivity of the design to the failure duration of control-system elements is determined. This includes the accuracy with which data on failure duration could be determined. In general, these data are assumptions only and therefore in the analysis the durations used were increased by a factor two, i.e.:

- one pump out of order: two days a year;
- all pumps out of order: one day a year;
- computer-system break-down: 10 days a year.

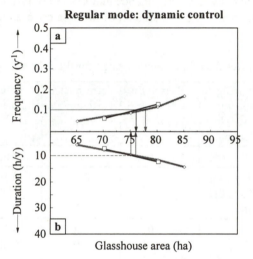

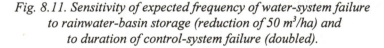

**Regular mode: dynamic control**

—□— Reduction of 50 m3/ha in rainwater-basin capacity
—▪— Doubled duration of control-system failure

*Fig. 8.11. Sensitivity of expected frequency of water-system failure
to rainwater-basin storage (reduction of 50 m³/ha) and
to duration of control-system failure (doubled).*

Figure 8.11 shows the results of the sensitivity analyses, performed for regular dynamic control. It is important to note that for both analyses the failure duration (graph b) is decisive if, in addition to the failure frequency, the criterion of failure duration is used.

When the actual capacity of rainwater basins is 450 m³/ha, the glasshouse area can be extended by 75 ha (Fig. 8.11) instead of 77 ha (Table 8.3). This shows that the design is not very sensitive to small errors in the size of the rainwater basins.

Doubling the duration of failures permits extension of the glasshouse surface to 76 ha (Fig. 8.11), slightly lower than obtained without doubling the duration.

Note that the sensitivity of the design to the parameters analyzed is greater when only the criterion of failure frequency is used (graph a in Fig. 8.11). Then the results are 76 ha and 78 ha respectively, instead of 80 ha (Table 8.3).

Given the sensitivity analysis, it can be concluded that, if both the failure-frequency and the failure-duration criteria are used, inaccuracies in the estimates of the parameters assessed, result in errors in the design of the order of magnitude of 2%. This error is of the order of magnitude of 4% when only the criterion of failure frequency is used.

On the basis of the sensitivity of the design to data inaccuracies, the designer has to decide whether extra capacities have to be included in the design. In the situation of the design example, which shows a small dependency on erratic data, a designer may decide to accept the design made without incorporating extra capacities. Otherwise, further iterations are required, to adjust the capacity parameters in the design and start with design step four again.

### 8.3.3  Advantages, Disadvantages and Improvements

*Advantages*

In comparison to the present standard design methods, one of the clear advantages of the dynamic design procedure is the possibility of including location-specific data. In addition to specific characteristics such as runoff, details on the operation of the water system can be incorporated in the design. Furthermore, the procedure includes the failure mechanisms that apply to regulating structures and the equipment used in water-system control. In the example described in the previous section, a trivial choice was made on this point, to include only failure of the central computer and the communication network in the dynamic control mode. Several other components can be included as well, e.g.: the availability of an on-line monitoring network; the maintenance situation of canals supplying water to pumping stations.

When the dynamic design procedure is followed to redesign a water system on the basis of water-quantity and/or water-quality variables, the excess capacities in the water system become known. When the requirements of interests change, these excess capacities can be used before the water-system has to be extended by, for instance, enlarging the regulating structures or surface-water storage.

Using the dynamic design procedure, excess capacities can be included on purpose. This approach may lead to designs that have similar properties as the ones found using the current standard methods. The advantage of the dynamic design is, however, that the locations and the extent of the excess capacities are known. This will probably give the responsible water authority more confidence in meeting the requirements of the interests present in the water system than the current practice using the standard design methods.

A very important advantage of the dynamic design procedure is that water-quality parameters can be included with relative ease. For example, the duration of violating water-quality standards can be incorporated, or the flushing factor needed to keep pollution in the water system below the pre-set limits. These options are not available in the standard designs.

The dynamic design procedure also enables determination of the periods that have the highest probability of failure during the year. This enables the water manager to carefully plan routine maintenance of regulating structures, minimizing the chance of system failure as a result of structures not being available at the moment they should be fully operational. Furthermore, other maintenance activities of the water system can be planned better. Excavation of silted-up canals, for instance, should take place before the full discharge capacity of the water system is likely to be required.

Summarizing, the dynamic design procedure produces a balanced and economic design, which includes the actual situation of the water system. The requirements of the interests are satisfied better and the actual safety present in the water system can be determined.

### Disadvantages

The standard design methods which are currently used, have the advantage that they are simple enough for a water system to be designed without many location-specific data. In this respect, the dynamic design procedure is more demanding. The result of the procedure depends on the data used. Therefore, sensitivity analysis is necessary. In case the data available are not detailed enough, the designer using the dynamic design procedure, will probably keep on the safe side and use data which influence the design so that excess capacities are again incorporated. As a result, the designed system capacities may be similar to the ones found applying the standard methods. In an extreme situation, the dynamic design may even include more capacity.

Depending on the water-system layout, the result of the dynamic design procedure can depend strongly on the probability of extreme system loads. If the critical failure frequency is 0.1 a year, using a time-series calculation of 30 years, the probability of three extreme events determining the water-system properties is relatively high. Each of these events, in principle, could have a less high actual frequency. For instance, one of the events could have a frequency of 0.001 a year.

Unfortunately, data are lacking to determine the frequency of extreme system loads more accurately. Simply using a series of 30 years, may therefore still result in some excess capacity in the design. It should be mentioned however, that all the current standard design methods have that same disadvantage, since these methods indirectly include the result of precipitation time-series analysis in the critical-discharge and the precipitation duration curves.

*Improvements*

As mentioned in the introduction of this chapter, the procedure described should be considered a step in the development to a more comprehensive form of water-system design. In practice, the method will probably require further improvement.

The case studies described in Chapter 7 showed that the process of calibration can be rather laborious and requires determination of many unknown system parameters. Therefore, one of the improvements, which is outside the scope of the present study but can be incorporated directly, is automatic calibration of models built in step two of the procedure.

Another improvement of the design procedure could be the use of condensed hydrological time series. If the design strongly depends on extreme system loads, a subset of the time series could be used to reduce the duration of calculations.

Condensation of time series is a very demanding job, however, since a high level of accuracy is required. The allowed level of condensation depends on the actual water system that the data are used for. The water system definitely has to be in an equilibrium situation that satisfies the requirements of all interests, before the reaction of the water system to an extreme event can be simulated.

## 8.4    Concluding Remarks

In this chapter both the standard design methods used and an alternative, the dynamic design, are presented.

A very important aspect of the dynamic design procedure is the possibility of including location-specific data, which is not possible using standard methods. Furthermore, dynamic design includes the dynamics of the water system, e.g. delays in runoff. The procedure enables including the possible control modes of regulating structures. This option permits incorporating the use of dynamic control in a water system at the design stage, taking advantage of all aspects of this control mode.

The example of the Delfland reference polder shows the difference between standard design and dynamic design. An important result is that the excess capacity in a water system designed by a standard method can be determined by means of the verification procedure. Probably, this is a general result of using location-specific data, since the standards used in the standard methods include accumulated safeties to fulfill the requirements, even in the most unfavorable situations.

The dynamic design procedure requires time-series calculations to include delays in runoff and incorporate the situation preceding system failure, such as saturation of the soil. The entire procedure is iterative and requires several calculations. For the design example, approximately 50 time-series calculations have been carried out for a period of 30 years. The procedure produces a lot of data, which have to be handled with care.

It is clear that the procedure described is only feasible if the entire process is automated, reducing the work of the designer to interpretation of the results and adjustment of system parameters for each iteration.

When the control mode of regulating structures is included in the design, the procedure yields the associated system capacities. If dynamic control is applied, its advantages in comparison to other control modes are automatically included, in fact, reducing the required system capacities.

In the example given, the total pumping capacity of the reference polder could be reduced by 14% using local automatic control and by 20% using dynamic automatic control. This result is obtained without any concessions to the failure frequency of 0.1 a year, nor to the failure duration of 10 hours a year.

If the total pumping capacity is left unaltered, the total area covered by glasshouses could be extended by 10% using local automatic control and by 28% using dynamic automatic control.

These results show that the responsible water authority can allow extensions in the water system for several years using local control, but for a much longer time using dynamic control. This means that investments in infrastructure, which seemed to be necessary, can be postponed for many years, while maintaining the required safety.

# 9 Conclusions and Recommendations

## 9.1 Conclusions

### 9.1.1 General

Current developments in water management force water authorities to control their water systems more accurately and take into account the requirements of a growing number of interests.

The current solution to changing requirements in water systems is to build extra capacity for storage, discharge or inlet of water. This thesis has presented a methodology that makes better use of the present infrastructure of a water system. If the methodology is applied, infrastructure scheduled can be postponed, reduced in size, or even canceled, while the necessary safety is maintained.

During the past few decades, local and central automatic control systems have been introduced for enhanced water management. Central control is executed by means of computers located in one place, which gather information from the water system and send control commands to the controllers of regulating structures. Typical for central and local control is that logic rules determine the control actions. Central and local control systems are widely applied to control subsystems of the water system. Dynamic control, a special form of central control, is the next logical evolutionary step.

In dynamic control, the entire water system, comprising rural and urban surface-water and groundwater subsystems, is considered. The interests distinguished, determine the way in which the water system should be controlled.

Methods similar to dynamic control have been studied and reported on by other authors. In the water-resources planning and management literature, storage and release problems in controlling entire river basins are frequently discussed. However, the problems described are generally of a nation-wide scale and require optimal control actions for periods of months to years.

In the present study, the critical period for which control actions have to be decided is hours to weeks. The subject resembles most real-time control studies carried out for sewer subsystems. However, in this thesis, sewer systems are considered only as subsystems of the water system.

To model and analyze regional water systems and the way in which these systems can be controlled, a Decision Support System (DSS) has been developed, called AQUARIUS. All water-system behavior results were obtained using the DSS in time-series calculations.

Much effort was paid to accurately model the various processes in a water system. Several models of water systems were created in practical case studies. The models were calibrated against water-quantity and water-quality-monitoring data. On the basis of these models, various control options were analyzed and evaluated.

Several control modes are distinguished, of which local manual control, local automatic control, dynamic manual control and dynamic automatic control were considered the most important ones.

Local automatic control generally improves water-system performance in comparison to manual control. Dynamic control, however, proves to be superior over local automatic control, especially when applied to complex water systems for which interests have been defined that have different and conflicting requirements regarding water quantity and water quality.

The automatic control modes require the regulating structures to be automated, which can be a costly solution, in particular when a water system contains many regulating structures. For that reason, special attention has been paid to combinations of control modes.

## 9.1.2  Water-System Capacity

Water systems basically lack the capacity to deal appropriately with extreme hydrological loads such as excessive precipitation or drought. Capacities of present-day water systems are determined on the basis of economics and frequency of system failure. It is generally accepted that some requirements of interests are sometimes violated, as long as the frequency is at a socially acceptable level. Depending on the type of interest, the frequency of failure can be several times a year to once every ten or more years.

Capacity allocation was only a matter of the design of the subsystems of a water system for many years. This thesis has proven that in the operational situation as well, some of the system capacities can be reallocated. This result could be achieved by two factors that are important in the operational situation: in general, not all interests need the capacity assigned to them in the design at the same moment and; water systems were found to have excess capacities, if controlled efficiently.

The requirements of interests are usually not static but depend on the season or even the time of day. When the capacity that is needed for one interest is not necessary at a particular moment, it can be used for another interest. Moreover, the hydrological load to subsystems of a water system generally satisfies or violates different interests under different conditions. For instance, the load resulting in failure of a rural subsystem differs from the load resulting in failure of an urban subsystem. Therefore, in this case as well, under extreme conditions, one subsystem can be operated to alleviate problems in another one. For instance, outside the growing season, agricultural land could be allowed to flood temporarily to prevent flooding in other areas.

In general, the capacities of the current water systems have been designed on the basis of general guidelines, which incorporate an unknown amount of safety. By means of time-series-calculations it has been proven that the excess capacities incorporated in the design, can be considerable. These excess capacities can be used in near-failure situations, to relieve interests which require extra capacity. For instance, the storage capacity in a sewer network can be entirely used before it overflows; a sewer overflow can be directed to a polder in which enough capacity exists to flush the surface water and effectively remove pollutants.

It is shown that system failures can be prevented or at least be reduced when the water system has sufficient unused capacity. Safeguards can be build to ensure that if a water system fails, the most important interests are violated last. These generally favor the common good, such as flood prevention and maintaining the ecosystem.

### 9.1.3 Control Strategy

In dynamic control the actions for regulating structures of the water system are determined for a control horizon. Coming hydrological events can be anticipated by taking control actions in advance. This is done by predicting the hydrological load to the water system, determining the associated system behavior and thus determining the optimal control strategy for the regulating structures of the water system. This is carried out by the Strategy Resolver of the decision-support level in the control system.

How to determine the optimal control strategy has been described by several authors, using various optimization methods. In general, the methods applied have in common that the strategy is determined by building and solving an optimization problem, which incorporates a model of both the water system and the objectives of control. That approach has serious drawbacks with respect to modeling accuracy in combination with the speed of finding the optimal solution. For complex water systems, to accurately solve the control problem will generally require nonlinear optimization methods that need long calculation times. If, additionally, the control horizon has to be incorporated in detail, which requires many time steps, it is questionable whether the process for finding the optimal control strategy is always faster than real time.

In contrast to the above methods, in the present study the functions of strategy determination and accurate water system modeling have been split up into separate simulation and optimization modules. This approach has the advantage that accurate modeling is possible in simulation, whereas in optimization the control problem can be solved on the basis of simplified relationships that describe the water-system behavior. This simplified model, however, still incorporates the processes of importance to water-system control, e.g. that of water-quantity and water-quality development.

The method developed is applied in two separate modules of the Strategy Resolver: a simulation module and an optimization module. The simulation module has the specific task of calculating present system states, on the basis of the nonlinear system behavior in detail, incorporating operational restrictions to control, such as regulating structures being out of order, limited capacities, etc.

The optimization module has the specific task to determine the optimal control strategy considering the requirements of water-related interests. In the methods developed, both modules work simultaneously, the simulation module continuously updating the current water-system conditions that are used in optimization.

The method proves to work very efficiently for regional water-system control in both polder and hilly areas. Practical operational conditions which limit the search space of optimization can be incorporated with relative ease.

A general distinction has been made between use of the method for the analysis of the operational situation and application in the operational situation itself. The use for analysis purposes is shown in this thesis. Several aspects of the application in the operational situation have been discussed in detail.


## 9.1.4  Optimization

### *Method Applied*

Several methods exist for solving the control problem and determining an optimal control strategy. Of the methods available these days, mathematical programming has been chosen mainly because it is universally applicable, includes deterministic modeling, is robust and makes model building and extension relatively easy. Several methods were reviewed from literature: Network Programming (NP), Linear Programming (LP), Successive Linear Programming (SLP), Dynamic Programming (DP) and Nonlinear Programming (NLP). The NP, LP, SLP and NLP methods were also examined within the framework of the research. After a careful evaluation of these mathematical programming methods, their characteristics and the objective of the present study, SLP has been chosen to solve the control problem defined.

The water-system model and the objectives of control are described by nonlinear relationships. In determining a control strategy for the control horizon, linearization on the basis of the present system state can be rather inaccurate, especially when long control horizons are used. Therefore, a method has been developed which converts the nonlinear problem into a linear one, but linearizes water-system processes accurately at the expected values of system variables in discrete time steps of the control horizon.

During ongoing time-series calculations, the accuracy of linearization is ensured by taking the optimal values found in each previous optimization cycle as the expected ones for linearizing. This method has been called forward estimating. It avoids multiple calculations of the same time step in a time-series calculation and has proven to yield accurate results.

The method of forward estimating requires linearizations for all nonlinear intrinsic water-system processes. The time needed to determine the linearization coefficients is small in comparison to the time taken to solve the optimization problem, especially for large problems. With respect to the speed of finding the optimal solution, the SLP method used is comparable to standard LP.

## Performance of the Method

Small or strongly generalized water systems, comprising only a few subsystems, a few flow elements, a limited number of interests and simple objectives of control, produce optimization problems containing around 1,000 variables and as many constraints. Large water systems comprising several interests, hundreds of subsystems, hundreds of flow elements and many objectives to meet, produce optimization problems containing 50,000 variables and as many constraints or even more.

Much attention has been paid to maintaining network structures and to scaling the problem in the application of the method in the DSS. The BPMPD solver used in the DSS is of the primal-dual interior-point type, which is rather sensitive to the sparsity of the constraint matrix and to whether a multi-diagonal structure in this matrix is maintained. When the optimization problem is built along these lines, it is very fast in finding the optimal solution. The efficiency in finding the optimal solution was furthermore kept high by scaling the problem well, keeping the differences in the order of magnitude of the matrix coefficients as small as possible.

Typical solution times obtained using a present-day regular 133 MHz Pentium PC with 48 MB internal memory, using an interior-point solving method, range from seconds for simple water systems to minutes for large and complex water systems. Such large and complex water systems cover areas in the order of magnitude of hundred thousands of hectares, modeled in great detail. These results were obtained when resolving optimal control strategies for control horizons with a length of 18 to 48 hours that cover a range of 12 to 36 time steps.

The solution speed of the method can be improved by reusing the optimal solution of each previous optimization cycle as the starting point in the optimization algorithm. Such speed improvements depend strongly on the type of solving method used. For the interior-point method applied, reductions in solution time of up to 50% could be reached.

The solution times found are so short in comparison to real time, that further improvement of the solution speed seems not really necessary at the moment from an operational point of view. This may change when extremely large models are built, comprising over ten thousands of subsystems, which may yield over a million variables and constraints in the optimization problem. However, computer power is increasing all the time as well.

## The Flow Problem

Several authors have described building optimization problems for sewer networks. The types of models used generally have in common that they only include continuity constraints to represent water transport in the optimization network. Delay in flow is incorporated by special time transportation arcs. The time transportation arcs allow delay in flow for only a multiple of the time step used in the optimization network. That approach can be inaccurate because only capacity restrictions can be included without the physical relationships of flow.

This thesis has shown that the use of time transportation arcs restricts the type of network that can be modeled. Looped networks are principally excluded from that type of modeling, which is a limitation in sewer-system modeling and definitely unacceptable in modeling regional water systems.

This thesis has presented a method of building the optimization networks which do not have the limitations listed above. A rectangular network for the optimization problem has been developed, which permits modeling loops in surface water, groundwater and sewer networks. In the approach, delays in flow are explicitly modeled by means of the physical friction relationships such as occur in water systems, modeling the various flow elements of a water system: regulating structures (i.e. pumping stations, controllable weirs, sluices, inlets), fixed structures (i.e. weirs, inlets) and free flow elements (i.e. canals, groundwater flow).

### *Difficulties Associated with Mathematical Programming*

A drawback of mathematical programming that is mentioned frequently is the difficulty to capture all interests in one objective function. The present study has proven that this is indeed difficult, but that it can be handled by using methods that incorporate the required nonlinear relationships in both the objective function and the constraints to the optimization problem. In the method developed, damage functions are defined for interests, which are not directly incorporated in the objective function but in separate constraints. These constraints are linked to the objective function via damage variables; each damage function having only one variable in the objective function. This has proven to be a very flexible method, which allows easy and multiple use of the same water-system variable in several damage functions.

Another point of general criticism to mathematical programming is the fact that costs or penalties should be assigned to undesirable situations. In this thesis it has been shown that this is not explicitly required, but that the same aspect can be incorporated in a mechanism which allows weighing of interests and inclusion of subjective policy preferences. Implicitly, the weighing mechanism uses numerical methods to determine penalties to undesirable situations. However, similar techniques were found to be necessary in the several weighing methods found in literature that include optimization.

### 9.1.5 Hydrological-Load Prediction

The hydrological data for off-line use of the DSS for analysis purposes and for on-line use in real-time control are different. For both types of use, the necessary hydrological data have been analyzed: forecasts from a weather bureau and predictions on the basis of 30-year monitoring data.

Verifications of forecasts show that they are generally inaccurate and very dependent on the type of weather, the time of year, the intensity predicted and the period the prediction is made for. Meteorologists of weather bureaus tend to underestimate excessive hydrological events.

A method has been developed to qualitatively assess the risk of control actions taken, taking into account incorrect forecasts. The risk-based approach presents the best control strategy to follow, regarding the uncertainty in the forecast and the possible undesirable results and damage to interests.

For analysis purposes, predictions by a weather bureau are simulated. Since for the past, no detailed forecast data were available for the periods required, use has been made of a hydrological database of past events. The data retrieved from the database have been disturbed and reduced, to simulate erratic predictions by a weather bureau.

Prediction of the hydrological load to a water system can play an important role in the determination of the optimal control strategy. For water systems that react slowly (e.g. mainly rural systems), it has been proven that only an indicative prediction of the load is needed to enhance water-system performance considerably. For water systems with rapid runoff characteristics (e.g. mainly urban systems), the suitability of the control strategy depends much more on the accuracy of the prediction available.

Prediction of hydrological variables by the weather bureau, are most accurate for short periods of time (e.g. hours), whereas these predictions are less accurate for long periods (e.g. days). Slowly reacting water systems require predictions for long periods and fast-reacting water systems for short periods. The accuracy of the hydrological-load prediction therefore seems to mirror the accuracy requirements. However, the accuracy required to effectively use the anticipating effect of dynamic control in fast-reacting water systems is still far from ideal.

Moreover, it was found to be better not to use a prediction at all than a highly incorrect one. Therefore, in real-time control the accuracy of the prediction should be supplied with the prediction itself by a weather bureau, which permits the water manager to exclude an uncertain predicted event from the data used for resolving the control strategy.

If no prediction is available, dynamic control uses at least the present system state to determine the best control strategy. Using the present system state in this way, is a kind of feedback in control. If the prediction of the hydrological load is moreover incorrect, the feedback guarantees that the control actions of the current moment are feasible considering the requirements of interests.

Even without a system-load prediction, the control strategy determined in dynamic control improves water-system performance in comparison to local control. The reason is that dynamic control on the basis of actual data, still reallocates capacities in the water system if possible, which is not possible using local control.

## 9.1.6 Interest Weighing

In water systems, three groups of interests can be distinguished: common-good interests, sectoral interests and operational water-management interests. These interests can be explicitly defined and mutually weighed, using a special weighing mechanism. The requirements of interests are linked to the physical water system by dimensionless damage functions. These represent the damage to interests as a function of water-quantity and water-quality variables. Damage to an interest occurs when the water-system variables of importance to that interest differ from their required values.

Weighing of interests is performed by decision makers in water management, on the basis of laws, government regulations, national and regional plans. Such weighing can never be entirely objective. Dynamic control requires careful assessment of the results. The results obtained should be presented to the decision makers for feedback. This is a cyclic and preferably interactive process, which confronts the decision makers with the choices they make and enables them to adjust their decisions, even before they are put into practice.

The weighing mechanism developed, shows that interests of different types can be compared to determine the best way to control a water system. Common-good interests can be considered successfully together with sectoral interests and operational water-management interests. The mechanism requires assigning the highest weights to the most important interests, trying not to violate the requirements of other interests.

### 9.1.7 Dynamic Water-System Design

At present, water systems are designed on the basis of guidelines concerning discharge capacity, surface-water storage and frequencies of extreme hydrological loads. These guidelines include generalized data on soil texture and land use. Time-series calculations for water systems designed on the basis of these guidelines, prove that the designs may incorporate a considerable excess capacity. The extent of this excess cannot be determined easily, however.

An alterative to the standard design method has been developed in the present study, which includes system dynamics and stochastics. The dynamic design procedure is carried out on the basis of a deterministic model of the water system, the possible failure mechanisms of control systems including regulating structures and standards with respect to the frequency of water-system failure. The dynamic design procedure presents an economic design method, in which excess capacities can be avoided. When required, however, excess capacity can be incorporated on purpose, to enable future extension of the water system.

The dynamic design procedure can be used to determine the safety of current water systems, on the basis of real data. In case a new water-system has to be designed, the procedure can be carried out on the basis of the requirements of water-related interests. In that case, violation of one of the requirements determines a water-system failure. This approach enables a balanced design in which all interests are incorporated, while the design explicitly takes into account how various requirements should be met. By doing so, in addition to water quantity standards, water quality standards can be included as well.

### 9.1.8 Practical Results

*General*

Practical case studies have been described in detail for the following regional water systems: Delfland (polder area), De Drie Ambachten (polder / hilly area) and Salland (hilly area).

The practical case studies show that water-system control in polder areas differs greatly from that in hilly areas. The span of control of regulating structures is generally wider in polder areas and these structures can generally control the water system both during the winter and the summer seasons. Despite the sometimes high number of weirs used in hilly areas, these structures alone cannot preserve enough water to satisfy the requirements of all interests, in general.

To assess potential improvements in water management, the performance index has been introduced, which is based on the extent of satisfying the requirements of interests. The performance index has proven to be an extremely helpful measure in comparing the long-term effects of alternative control modes.

To assess the results of alternative control modes under extreme conditions, failure situations have been defined for the most important interests. A failure in this respect is defined as the situation in which one requirement of an interest is violated, e.g. flooding, fish mortality, harvest reduction. Failure frequencies were determined in time-series calculations by recording the number of times the variables that determine whether the requirements are met, exceeded the limits of their permissible range.

### Polder Areas

The polder areas studied show a great improvement in water-system performance in general. Anticipating high hydrological loads is of importance in some of these areas to prevent water-system failure. In those cases, the hydrological-load prediction should be an element of the control system.

Various examples show that the control strategies determined are not very sensitive to the accuracy of hydrological-load predictions. In mainly rural areas, with slow runoff characteristics, the results of determining the control strategy without prediction are almost as good as with prediction. However, in urban areas prediction can be necessary to improve water-system control.

In the polder areas investigated, the number of failures following excessive precipitation could be reduced considerably by dynamic control. This effectively prepares the current water-system arrangements to better deal with new interest requirements when dynamic control is applied.

### Hilly Areas

The improvement in water-system control in the hilly areas is less great than in polder areas. The main reason is that the requirement to keep the groundwater table as high as possible in summer, generally prevails. It proves to be very difficult to prevent a fall in groundwater level using weirs only, because the fall results from uncontrolled processes such as evaporation and infiltration into low-lying groundwater aquifers. For that reason water supply works or at least retention measures are usually required.

In general, the best possible improvement in groundwater preservation is achieved by keeping the water levels in the adjacent surface-water subsystems as high as possible throughout the year and lowering them only when excessive runoff can be expected. Part of this strategy can be accomplished by automatically controlled weirs.

However, in some places the permeability of the soil is so high that surface water upstream from a weir infiltrates into the soil and flows through the subsurface as shallow seepage into the surface waters downstream. This effect, combined with infiltration throughout the year, often necessitates water supply via pumping stations. In hilly areas, the supply of water generally has to be carried out in small stages and therefore small, but complex systems have to be developed.

In the hilly areas investigated, a rise in groundwater level of 0.5 to 1 meter could be achieved if artificial water supply is combined with dynamic control. Furthermore, drought problems could be better handled.

### Water-System Performance

The examples given in this thesis show that dynamic control can considerably enhance water-system performance and that it is superior to local control in various ways. In the discussion whether to apply dynamic control or not, the actual performance improvement should be assessed. This can only be done by means of dynamic time-series calculations that determine how the water system reacts when dynamic control is applied.

In water systems that include a large number of uncontrolled structures, the application of dynamic control may imply automation of a large part of these structures, which can be costly. The choice which structures to include in dynamic control is therefore partly an economic one, which should be evaluated at the feasibility stage of a design. However, the costs of automation equipment are generally much lower than that of building extra capacity in the water system by extending the discharging capacity or excavating canals of the surface-water system.

The case studies show that it is generally not necessary to automate all regulating structures in a water system. Moreover, not all automated structures have to be controlled dynamically to obtain good results in operational water management. Only the structures which play a key role in preventing damage to interests, have to be included in a dynamic control system.

### 9.1.9  Application of Dynamic Control

### Decision Support

As mentioned before, the methodology implemented in the DSS as presented, can be applied in two possible ways: off-line, for water-system analysis and on-line, for real-time control.

Off-line application of a DSS is of importance to assess the potential improvement which can be obtained by using various optional control modes. This application is also of interest in the planning stage of a real-time control system, because it helps determine which regulating structures should be remotely controlled, locally controlled or maybe not controlled at all. In addition, off-line application of a DSS can be used in operator training, simulating known and unknown extreme hydrological conditions. Moreover, irregular conditions can be simulated in which the operators have only limited means of control, such as during maintenance or predictable system-failure situations.

On-line application is implemented in a remote control system. The on-line DSS determines a control strategy on the basis of actual data. It runs on a centrally placed computer, that receives filtered monitoring data from the water system and sends the control actions required to the local control units.

Monitoring data from the field cannot be used directly. A numerical filter has to be built into the control system to adjust the simulation model in such a way that it continuously resembles real-time water-system behavior. This filter can also be used to determine data that are measured with a low frequency.

Dynamic control generally only has to come into action when the system capacity during the control horizon becomes too low and/or interests in the water system are conflicting. In water-quantity control, this usually only occurs during high hydrological loads. When dynamic control is applied, alternative, less advanced control modes should also be available, e.g. local control. These control modes have to come into action in case part of the central control system breaks down.

### *Implementation*

Dynamic control has been presented as a method which incorporates weighing of interests in the entire water system. Regional water systems are generally controlled by various water authorities, which requires weighing of interests that override the boundaries of responsibility of one authority. Because each water authority has its own responsibilities, such weighing has not been carried out in the operational situation so far. Consequently, each water authority focuses mainly on his own subsystem. This policy can only be changed if it is proven that using the entire water-system state to determine control strategies really improves the situation in the regions controlled by all authorities.

To persuade water authorities, the general advantages and improvements in control have to be demonstrated. However, in practice many administrative hurdles still have to be taken.

When implementing dynamic control, the experience and knowledge of operators will continue to be of vital importance to successful operation of water systems. In special situations such as an excessive hydrological load, it will be the operator who has to supervise the automatically executed control commands. If necessary, he will have to take responsibility, possibly taking over manually and steering the processes in the water system himself. The knowledge of operators can be further enhanced by use of the off-line DSS as mentioned above, making them familiar with special situations in a short period. To gain this type of experience in the operational situation would take many years.

## 9.2    Recommendations

The subject of optimal control of water resources systems has been studied by researchers in several countries. Many questions remain if one wants to further develop methods to enhance water-system performance. The present study has solved several of the problems involved in optimal control of a regional water system. However, many topics remain which were outside the scope of this study. The most important topics for further research are summarized below.

### 9.2.1  Optimization

*Accuracy*

The accuracy of the solution to the optimization problem depends on the size of the time step used in the control horizon. In the performance analysis of the method developed, it was shown that the time to solve the optimization problem is generally more or less a power function of the size of that problem. The problem size depends linearly on the number of time steps used in the control horizon. Therefore, the control problem is solved most rapidly when the time step used is large. However, to keep on the safe side, relatively small time steps have been used in the case studies presented in this thesis.

Calculating the 30-year time-series for the case studies presented in this thesis, took hours to days. To speed up the overall time required to calculate time series, it is recommended to enhance the method and determine the time step required during the calculation automatically. This can be done by means of an automatic sensitivity analysis of the solution to the size of the time step during the ongoing process. A criterion for reducing the time step could be the total damage determined by optimization during the control horizon. If the damage is outside a pre-set range around the total damage determined in a prior time step, the size of the time step is reduced, if possible.

An option to further enhance the accuracy of the forward-estimating method, is to include an automatic evaluation procedure, which determines whether it is necessary to perform a second forward-estimating cycle at the current time step in a time-series calculation. The proposed evaluation determines, for a selected set of optimization variables, whether the values of these variables at a specific time, are outside a range around their values in a previous cycle. If one of the variables is outside the range, another forward-estimating cycle should be performed at the same time step, using the outcome of the prior cycle to determine the linearizations of processes in the water system more accurately. If the selected variables fulfill the evaluation criteria, the method proceeds to the next time step in the time-series calculation.

It is expected that, when the method of simultaneous simulation and optimization is extended with the above-mentioned use of large time steps whenever possible and the automatic accuracy evaluation of the forward-estimating method, the total time to calculate long time series can be reduced, without making concessions to the accuracy of the method.

*Alternative Method*

The optimization method applied in this thesis is a form of nonpreemptive goal programming. The method does not include an explicit hierarchy in objective functions.

It is recommended to study the possibilities of preemptive goal programming to solve the optimization problem defined. Preemptive goal programming allows incorporation of a hierarchy in objectives. The optimization problem is first solved trying to meet only first-priority objectives. The remaining solution space is used to meet second-priority objectives, etc. This approach allows modeling of a hierarchy which usually can be distinguished for interests, i.e.: common-good interests, sectoral interests and operational interests.

*Other Applications*

Despite the excellent reviews of mathematical programming methods applied in the field of water resources management, published by several authors, the choice for a particular type of method still seems to depend on the field of use, the country of application and the preference of the modeler.

Within the scope of the present study several methods were reviewed, from which the Successive Linear Programming (SLP) method seemed to have the greatest potential for water-system control in the regional water systems studied. The application of SLP presented here, can be considered a competitive alternative to Nonlinear Programming, especially with respect to the size of the problem which can be solved and the speed of finding the optimal solution.

It is recommended to further investigate the possibilities of SLP in combination with forward estimating in the field of water resources management. A special point of interest is whether the method developed can accurately include the interest of generating hydropower and the process of the associated optimal reservoir releases in large river basins.

In addition, the method developed could be of interest to other fields in which flow problems occur. It is recommended to investigate the possibilities of its use in industry such as gas transport where currently network programming methods are frequently used, but also in oil and electricity transport.

## 9.2.2  Interest Weighing

The way in which interests are weighed and used in optimization to determine the optimal control strategy, is incorporated explicitly in the methodology developed here. Preferences of decision makers can be made visible and the effects of these preferences can be assessed by means of the DSS. The approach chosen is to determine the weights of interest iteratively and preferably by an interactive process together with the decision makers in a water agency. The approach focuses on the required system behavior, rather than on fixing target values and standards for water system variables. In the present research, the approach has been tested in practice on a limited scale only.

A separate study of the process of interactively finding the required water system behavior is recommended, involving a group of decision makers who discuss and balance interests. The setting can be similar to the one used currently, to fix target values and standards in water-management organizations. The purpose is, to find the operational flexibility in water systems, to determine the conditions required during average system loads and the system behavior in failure and near-failure situations and to incorporate the specific requirements of different interest groups in operational control.

### 9.2.3 Hydrological Loads

A first attempt has been made to develop a risk-based approach to determine which hydrological load has to be accounted for in water-system control. In the approach, the risk of using incorrect hydrological predictions, is incorporated.

It is recommended to further study the risk-based approach, so that forecasts of weather bureaus can be handled and used with confidence in real-time control of water systems. The damage functions presented in this thesis can be used in that case to assess the effect of incorrect predictions.

It is furthermore recommended to study the alternative of chance-constrained programming, which includes the uncertainty of the hydrological load by imposing probability distributions of loads on the optimization problem.

### 9.2.4 Process Descriptions

The flow-process descriptions applied in this thesis present satisfactory results in modeling for regional water-system control. It has not been the primary purpose of this study to describe all flow processes in detail and therefore some descriptions are very general and are open to improvement.

It is recommended to more accurately describe the discharge behavior of a spilling sluice. The various flow conditions have to be determined and simulated in detail, with emphasis on the transitions between these conditions and associated hysteresis effects.

Furthermore, the way in which surface-water flow is modeled in simulation and optimization could be improved by, for instance, using the kinematic wave equation. This permits delays in flow to be incorporated more accurately.

Both improvements can be implemented without fundamental changes to the methodology.

### 9.2.5 Water Preservation

Extensive drainage in hilly areas has resulted in large-scale lowering of the groundwater table. Water sometimes runs off from these areas, before it can recharge the groundwater. Because of shallow seepage effects, building weirs in water courses is generally not sufficient. To solve the problem, in principle, surface-water discharge via canals could be restricted. However, these canals were originally intended to prevent flooding in times of excessive precipitation.

Several options to improve water preservation can be distinguished. The option which is recommended to be studied further is the use of shallow and wide canals. Special attention should be given to the problem that shallow-water vegetation will develop faster, which increases canal bed roughness, which restricts the discharge capacity and/or may necessitate more maintenance.

### 9.2.6  Determining Logic Rules

Dynamic control as shown in this thesis uses optimization to determine control strategies for water systems. The possibilities of deriving general rules from the control strategies determined, has not been studied. However, it is believed that for many water systems, especially simple ones, this can be a feasible alternative to on-line use of the DSS.

Therefore, it is recommended to carry out a study to automatically determine logic rules for water-system control on the basis of optimal control strategies. As shown in this thesis, several options are available for on-line use of such logic rules. One of the options is the verification method, which has the advantage above others that the process by which a control strategy is determined can be monitored easily by the operators.

### 9.2.7  Filtering of Monitoring Data

One of the reasons why optimization-based control systems have not always been successful in practice, is that the models used to determine the optimal control strategy, do not always represent the actual water-system state. When a discrepancy occurs between model and practice, it cannot be guaranteed that the solutions found in optimization indeed reflect the optimal control of the actual water system. Therefore, in the implementation of these advanced methods, it should be assured that the modeled water system closely resembles reality all the time. This can be done by incorporating data on the availability of regulating-structure capacities, but also by using filtering methods that incorporate real-time monitoring data in the models used.

It is recommended to enhance the functionality of AQUARIUS, the DSS developed as part of this study, with a numerical filter. The control strategies provided by the methodology presented in this thesis can then be implemented on line.

### 9.2.8  Designing with Dynamics

In this thesis an iterative verification procedure is presented to determine optimal water-system design, using the dynamics of interests, the water system, the hydrological load and failure mechanisms of regulating structures.

It is recommended to further enhance the method and apply it to more complex regional water systems. Of special interest in the further development of the method is automatic calibration of models and automatic data handling.

As an alternative to the dynamic design procedure, pure stochastical methods such as the Monte-Carlo analysis can be considered. Such analyses generally need very long time series for calculation, since a reasonable number of failure situations must have occurred together before conclusions can be drawn. This can be seen as a special point of interest to study, especially for water-system modeling, in which previous conditions in the system play an important role in the occurrence of a failure.

### 9.2.9  Use of a DSS by Operators

Using a DSS in regional water-system control is a relatively new development, which has to be introduced gradually. Operators should be trained in the use of such systems. The outcome of the DSS may not always look very logical. Therefore, a prerequisite to successful implementation of these systems is intensive training, not only to obtain knowledge on how the system works, but also to gain confidence in the outcomes.

In addition, it is very important to involve the actual operators, as early as the design stage, in the building of a real-time-control system that includes a DSS. Their experience ensures that practical constraints, which are not detected during desk studies, are incorporated in the design.

In general, it will be the operator who has the practical responsibility for the operation of a water system. The question arises which action he should take when the DSS presents unexpected results: follow the advice of the computer or his own intuition and experience.

When using a remote control system, every operation by an operator is monitored and stored in a database. This makes it easy to trace incorrect decisions afterwards, an aspect that is generally absent in traditionally controlled water systems.

It is recommended to further study these subjects, also with respect to the legal consequences. More than before, operators will need clear guidelines to be sure that they take the right decision and operate the water system according to the regulations.

# Glossary

**Abstraction area**
The *area* from where water is abstracted by a *flow element* of a water system.

**Alien water**
Surface water from outside the water system under consideration.

**AQUARIUS**
*Decision Support System* for water-system control. AQUARIUS comprises a set of computer programs especially developed for water-system modeling, analysis and real-time control. Models in AQUARIUS are build by configuring schematic representations of water systems in a network layout, using various *water-system elements*. AQUARIUS can be used for time-series calculations to determine the water-system performance, using various control modes. The DSS enables definition of interests and damage functions and, during its time-series calculations, determines failure frequency and failure duration for specific water-system variables. AQUARIUS includes the *Strategy Resolver* which can determine optimal control strategies and *dynamic control* of water systems.

**Arc**
Schematized *flow element* in an *optimization problem* which connects two *nodes* of a network and represents flow ('transportation arc') or storage ('storage arc') (Fig. 6.8, Page 118).
Two types of transportation arcs exist: 'time transportation arcs' and 'time-independent transportation arcs'. Time transportation arcs connect nodes belonging to different locations in a water system at different time steps and can be used to model delay in flow, without having to incorporate the intrinsic relationships of flow. Time-independent transportation arcs denote water transport from one location to another in the same time step. When the latter are used, the intrinsic relationships of flow have to be included in the formulation of the problem, to simulate delay in flow. Time transportation arcs cannot be used to model looped networks, whereas time-independent transportation arcs can.

**Area**
*Water-system element* comprising *subsystems* for which processes such as runoff, water use and water quality development can be simulated.

**Boundary**
*Water-system element* describing a natural or an artificial separation of *water systems* or *subsystems* of a water system. In modeling, boundaries are included to enable detailed simulation of the behavior of a water system, without having to describe the entire water system and its environment.

**Central control**
See *control level* and *control mode*.

**Common cause event**
An event which causes an entire unit of a control system to fail, e.g. lightning striking the electricity supply network, the consequence being that all pumps of a pumping station fail to operate.

**Common-good interest**
A common-good interest includes aspects such as whether the land is fit for habitation and focuses on improvement and preservation of the environment. A common-good interest entails requirements related to primary water-management duties. This *interest type* is generally of importance to an entire water system.

**Communication network**
Control system *component*, consisting of telephone lines, direct connected lines or other communication lines, including the equipment that enables data communication.

**Component of a control system**
Hardware part of a *control system*, e.g. a computer, a data communication line.

**Constraint**
Physical or artificial limiting condition to an *optimization problem* (Eq. 3.1, Page 45). Two types of constraints are distinguished: *continuity constraints* and *side constraints*.

**Constraint Manager**
Part of the *optimization module* that converts particular water-system relationships in combination with the *water-system state* and the hydrological-load prediction into an *optimization problem* (Fig. 6.1, Page 110).

**Continuity constraint**
*Constraint* describing continuity for water quantity or water quality in a *subsystem* of a modeled water system.

**Control horizon**
The period for which a *control strategy* has to be determined.

**Control level**
Central control: control of one or more *regulating structures* on the basis of data from more than one location.

Local control: control of one regulating structure on the basis of monitoring data gathered in the vicinity of that structure.

## Control mode

The following six control modes are distinguished (Table 2.1, Page 30):

Dynamic Automatic Control / Dynamic Manual Control;

Central Automatic Control / Local Automatic Control;

Central Manual Control / Local Manual Control.

In dynamic control mode, operations are based on the time-varying requirements of interests considered in a water system, the water-system load and the dynamic processes in the water system.

In central and local automatic control mode, operations are based on logic rules, e.g. a decision tree.

In manual control mode, operations are based on heuristics, which may include experience, knowledge and reasoning.

## Control strategy

A sequence of control actions for some time ahead, which, given a predicted water-system load, result in the required system behavior (Sec. 2.3.1, Page 27). The control strategy is determined for the duration of the *control horizon*. It can be determined by trial and error, using simulation, or in a single run, using optimization. The control strategy can be formulated in the form of a sequence of *setpoints* in time.

## Control system (In Dutch: 'besturingssysteem')

A computer-based system which controls *regulating structures* in a water system. A control system consists of one or more computers and a *communication network*. Operation of these components depends on the type and implementation of the control software.

## Control variable

Variable used to control a water system, e.g.: water transport through a pumping station.

## Damage

The dimensionless product of a *penalty* coefficient and an *optimization variable*, which expresses the weighted harm to one or more interests defined for the water system.

## Damage function

Function which schematically expresses the damage to a water-system interest, depending on the state of a *key variable*. The damage to one interest can be determined by several damage functions.

## De Bilt time series

Hourly precipitation and daily evaporation measurements as determined by the KNMI in the 30-year period of 1965 to 1994, at the De Bilt meteorological station.

**Decision Support System** (DSS)
An interactive computer-based system that helps decision makers to use data and models to solve complex problems (after Sprague & Carlson, 1982).

**Delivery area**
The *area* into which water from a *flow element* of a water-system is discharged.

**Dimensionality**
A measure that relates computer resources of a *solver* to the size of an *optimization problem* and its level of nonlinearity, nonconvexity and discontinuity, in which the size of the problem is measured by the number of *optimization variables* and the number of *constraints*.

**Dynamic control**
See *control mode*.

**Dynamic-control threshold**
The value of the total *damage* to the water system at particular time step of the *control horizon*, at which local automatic control should switch over to dynamic automatic control.

**Dynamic Design**
Procedure based on verification of the performance of a particular water-system, using time-series calculations that determine the *failure* frequency and duration of the water system. The dynamic design procedure takes into account the layout of the water system and the actual conditions of soils, waterways, and regulating structures.

**Failure**
The undesirable situation which occurs when the requirements of one or more of the interests involved in a water system are violated in such a way that the objectives set cannot be met, e.g.: flooding, fish mortality, loss of harvest. If one or more requirements of an interest fail, the interest fails and consequently the *water system* fails. This is called a 'system failure'.

**Feasible solution**
Valid solution to an *optimization problem*.

**Fixed structure**
*Flow element* which cannot be controlled or can be controlled manually, but very infrequently, e.g.: a fixed weir, an inlet which is opened during summer and closed again for the winter season.

**Flow element**
*Water-system element*. Flow elements can be divided into controllable flow elements, called *regulating structures*, fixed flow elements, called 'fixed structures' and 'free flow elements'.

**Flushing factor**
The flushing factor gives the ratio of water which is flushed through the water system each day for water-quality control, and the surface-water volume at target level (Eq. 7.2, Page 165).

**Forward estimating**
Method to avoid multiple calculations of the same time step in a time-series calculation by the *optimization module*. The method ensures accurate linearization of the nonlinear water-system relationships used in SLP. Linearizations are determined at the values expected for *water-system variables*. For each time step, the expected values are abstracted from the optimal solution found in each previous step calculated (Sec. 6.2.2, Page 111).

**Free flow element**
*Flow element* which cannot be controlled and of which the flow is determined entirely by upstream and downstream water or pressure levels, e.g. a canal; groundwater flow.

**Generalized model**
Model of a water system which describes the hydrological processes that are important at a general scale, but is still accurate enough to represent the water system.

**Global optimum**
The overall optimal solution to an *optimization problem* (Fig. 3.1, Page 41).

**Groundwater flow**
Saturated flow through the soil, resulting from a piezometric pressure difference in the soil. Shallow groundwater flow: phreatic and mainly horizontal groundwater flow. Deep groundwater flow: mainly horizontal flow through aquifers and vertical flow through aquitards.

**High water-system load**
*Water-system load* that results in undesirable effects, which cannot be prevented by the operation of regulating structures, e.g.: a continuous and undesirable water-level rise in a subsystem, while the entire drainage capacity of that subsystem is in use.

**Hilly areas**
Gently sloping areas, generally drained by gravity.

**Hydrological load**
*Water-system load* consisting of precipitation and evaporation.

**Infeasible problem**
An *optimization problem* to which no solution can be found by the *solver* applied.

**Integrated model**
A dynamic simulation model or models which describe all the relevant processes and control actions in the various subsystems of a water system.

**Interest weighing**
Mechanism to assign priorities to the various interests present in a water system (Sec. 1.3.5, Page 18 and Sec. 1.3.5, Page 18)

**Interest type**
Three types of interests are distinguished: *common-good interests*, *sectoral interests* and *operational interests*.

**Key variable**
*Water-system variable*, used in a *damage function* to represent the level of satisfaction of interest requirements.

**Local control**
See *control mode*.

**Local optimum**
A solution to an *optimization problem* which is only optimal within a limited range of the available solution space (Fig. 3.1, Page 41).

**Logic controller**
Computer system which, on the basis of the actual *water-system state* and logic rules, determines the required *control strategy*.

**Manual control**
See *control mode*.

**Mathematical optimization**
Category of deterministic optimization methods, using a mathematical problem formulation, e.g.: Network Programming, Linear Programming, Successive Linear Programming, Dynamic Programming, Nonlinear Programming.

**Milling stop** (In Dutch: 'maalstop')
A milling stop is imposed when the water level in a *storage basin* is very high. Then some, or all polder pumping stations are no longer permitted to discharge water to that storage basin. The reasons for water board deciding on a milling stop are e.g.: to prevent flooding of the hinterland or damage to canal embankments.

**Modeling element**
Part of a model, here representing *areas*, *boundaries* and *flow elements*.

**Module**
Computer program or part of it.

## Network-flow model

A model that only consists of *flow elements*. In a network-flow model all flows and storages are represented by *arcs*. The *nodes* in a network-flow model solely represent places where arcs intersect.

## Node

Network intersection which represents a location for water storage. The use of 'supply nodes' at the boundaries of a network enables modeling of inflow from outside the network. The use of 'demand nodes' at the boundaries of a network enables modeling of outflow from the network to the surrounding area.

Furthermore, 'basic nodes' and 'time nodes' are distinguished. Basic nodes in a network represent the original locations in a water system and form the basis of a network that expands in time. Time nodes in a network are replications of basic nodes at discrete time steps of the *control horizon*.

## Nonpreemptive Goal Programming

Mathematical *optimization method* in which, for each objective, a specific numerical goal is set and an *objective function* is defined. The optimal solution is found by minimizing the weighted sum of deviations of objective function values from their respective goals.

## Objective function

A function of *optimization variables* that presents a measure to evaluate the satisfaction of objectives set to an *optimization problem*.

## On-line monitoring

Monitoring by a monitoring network which offers automatic data gathering and transmission.

## Operational interest

An operational interest includes issues that determine efficient water-system control. This *interest type* is generally only important for the responsible water authority and requires the best operational control at the lowest cost.

## Optimization

To select the best possible decision for a given set of circumstances without having to enumerate all the possibilities (Pike, 1986).

## Optimization method

A method that can be applied to solve an *optimization problem*.

## Optimization module

*Module* consisting of the *Constraint Manager* and a *Solver* (Fig. 6.1, Page 110).

## Optimization problem

Mathematical model formulation which consists of one or more *objective functions* and *constraints*.

**Optimization variable**
*Water-system variable* included in an *optimization problem*.

**Parameter**
Fixed value which represents a system characteristic, e.g.: the hydraulic conductivity of the soil; the discharge coefficient of a weir.

**Penalty**
Coefficient in the *objective function* that assigns a damage to an *optimization variable*. An alternative term for this coefficient, which is also frequently used in literature is 'cost coefficient'.

**Performance**
Term used to indicate the extent to which a controlled water system achieves the objectives which have been set for it.

**Performance index**
The total sum of damages in a water system over time in a reference situation, divided by the total sum of damages when using an alternative set of control modes (Eq. 7.1, Page 163).

**Polder**
A level area which was originally subject to a high water level, either permanently or seasonally and due to either groundwater or surface water. An area becomes a polder when it is separated from the surrounding hydrological regime so that its water level can be controlled independently of the surrounding regime (Segeren, 1983).

**Polder areas**
Flat areas consisting for the greater part of *polders*.

**Prediction module**
*Module* which determines the *water-system load* during the *control horizon* on the basis of real-time and historical monitoring data (Fig. 6.1, Page 110).

**Preferred range**
The range within which a *water-system variable* should be kept to best satisfy a specific interest in the water system. This range can vary with time (Fig. 7.1, Page 157).

**Preservation factor**
The ratio of the quantity of water used and the total quantity abstracted from the system. The abstracted quantity is the sum of: water flowing out of the water system (e.g. through flow elements) and the quantity used up by the interests in the water system (e.g. total evapotranspiration) (Eq. 7.3, Page 165).

**Real-time control** (RTC)
Control on the basis of real-time monitoring data, in which the time lapse between measurement and control action is short in comparison to the response time of the controlled system.

**Reclaimed lake** (In Dutch: 'droogmakerij')
Low-lying *polder*, which has been reclaimed by draining a lake, using pumping stations.

**Regulating structure**
Structure which is regularly operated manually or automatically, adjusting its flow capacity. Examples of regulating structures are: pumping stations, controlled weirs, sluices and controlled inlets.

**Remote control**
Control commands given from a distant location, e.g. a *regulating structure* can be controlled remotely from a central location.

**Saturated zone**
The zone in the soil below the groundwater table.

**Seasons**
In this thesis: winter (1 January - 31 March), spring (1 April - 30 June), summer (1 July - 30 September), autumn (1 October - 31 December). Spring and summer together are also referred to as the 'summer season', while autumn and winter together are referred to as the 'winter season'.

**Sectoral interest**
A sectoral interest is characterized by the benefits that a particular group derives when its requirements are met. A sectoral interest involves allocation of water to a specific activity. In a sectoral interest, socio-economic balancing and/or social acceptance can be involved. This *interest type* is generally of local importance.

**Setpoint**
The target value for a variable or limit to a variable (Sec. 2.3.1, Page 27).

**Side constraint**
*Constraint* in an *optimization problem*, used for describing a physical flow relationship (not being a continuity equation), e.g. the flow equation for a weir.

**Simulation module**
*Module* which includes deterministic process relationships that describe the nonlinear water-system behavior (Fig. 6.1, Page 110).

**Simulator**
A running calibrated simulation model.

**Simultaneous simulation and optimization**
The technique of simulation and optimization program modules of the *Strategy Resolver* working together and at the same time transferring data.

**Soil moisture content**
The volumetric fraction of water in the soil.

**Solver**
Computer program which applies a specific algorithm to find the optimal solution to an *optimization problem*.

**State variable**
Variable which is used to describe a *water-system state*, e.g.: a water level, pollutant concentration, a soil moisture content.

**Storage basin** (In Dutch: boezem)
System of interconnected water courses (ditches, canals, lakes, etc.) into which the water of predominantly polders is discharged, serving as temporary storage and for discharge onto external waters. A storage basin can generally also be used to supply water from outside a water system to polders.

**Strategy Resolver**
System of simultaneously running *simulation*, *prediction* and *optimization* modules that serve to support the water manager in determining the best control strategy (Fig. 6.1, Page 110).

**Subsystem**
Part of a *water system* with distinctive location and/or runoff characteristics. A water system consists of *subsystems*, such as surface-water subsystems, groundwater subsystems and sewer subsystems.

**Successive Linear Programming** (SLP)
Special implementation of Linear Programming (LP), which enables accurate solving of nonlinear *optimization problems*. Successive Linear Programming, in its basic form, is an iterative procedure, in which the optimal solution of each iteration is used to determine the linearizations of nonlinear process descriptions in the definition of the next optimization problem more accurately. Iteration stops as soon as the values at which linearizing is done are close enough to the values found in the optimal solution.

**System failure**
See *failure*.

**Target level**
The required level of a water-system variable, e.g.: groundwater target level; surface-water target level.

**Uncertainty multiplier**
Multiplication factor applied to hydrological variables retrieved from an historical database to simulate forecasts by weather bureaus for specific periods of the *control horizon*. The uncertainty multiplier manipulates and generally reduces the stored values and thus achieves a more realistic prediction of the hydrological load on a water system than the original measured data.

**Unsaturated zone**
The zone in the soil between the soil surface and the groundwater table.

**Warm start**
Use of a previous solution to an *optimization problem* as a starting point to find the next solution. This option can be used by some *solvers* to speed up the process of finding the optimal solution. In general, the previous solution used should be close to the optimum to really speed up solution time.

**Water authority / water manager**
Two terms to indicate the organization responsible for water management. The term 'water authority' denotes the organization which is formally responsible. The term 'water manager' implies the person or group of persons within the organization who implement control actions.

**Water preservation**
Keeping as much water in a water system as possible, preventing drought problems and/or reducing the necessity of letting in *alien water*. See also *preservation factor*.

**Water system**
The entire water-related environment, consisting of open water, beds of water courses, banks, technical infrastructure, biological components and groundwater, which together form a hydrological unit.

**Water-system approach** (In Dutch: 'watersysteembenadering')
The approach which focuses on the coherent elements of a water system and the various interactions between the water system and its environment.

**Water-system element**
Model naming convention for *areas*, *boundaries* and *flow elements*.

**Water-system load**
Load to a water system, consisting of uncontrolled amounts of water, possibly carrying pollutants, which enter or leave the system, e.g.: precipitation, evaporation, seepage, infiltration, lock water, sewage water, STP effluents. See also *high water-system load*.

**Water-system state**
Set of values for *water-system variables* at a specific moment.

**Water-system variable**
Variable which presents a (temporary) situation in a water system. Two types of water system variables can be distinguished: *state variables* and *control variables*.

# Abbreviations

| | |
|---|---|
| BOD | Biochemical Oxygen Demand |
| BPMPD | Primal Dual Interior Point Method |
| CSO | Combined Sewer Overflow |
| DP | Dynamic Programming |
| DSS | Decision Support System |
| DWF | Dry Weather Flow |
| GWL | Groundwater Level |
| ISO | Improved Sewer Overflow |
| KNMI | Royal Netherlands Meteorological Institute |
| LP | Linear Programming |
| MIP | Mixed Integer Linear Programming |
| MPS | Mathematical Programming Standard |
| MSL | Mean Sea Level |
| NAP | Amsterdam Ordnance Datum (reference level, close to MSL) |
| NLP | Nonlinear Programming |
| NP | Network Programming |
| PI | Performance Index |
| PID | Proportional Integral Differential |
| PLC | Programmable Logic Controller |
| Ref | Reference level |
| RTC | Real-Time Control |
| SCADA | Supervisory Control and Data Acquisition |
| SLP | Successive Linear Programming |
| SSD | Separate Sewer Discharge |
| STP | Sewage Treatment Plant |

# Symbols

Only unique symbols, which are frequently used in this thesis, are listed below.

$\forall$ : for all
$\alpha_g$ : certainty parameter for direction of groundwater flow (-)
$\gamma_c$ : canal flow reversion factor (-)
$\eta$ : damage variable (-)
$\theta$ : soil moisture content (-)
$\kappa_{sl}$ : binary parameter to prevent backflow through a sluice (-)
$\kappa_w$ : binary weir parameter to prevent back flow (-)
$\kappa_{ws}$ : binary weir parameter for submerged flow (-)
$\mu$ : effective storage coefficient (-)
$\mu_{sl}$ : contraction coefficient for sluice gates (-)
$\tau$ : simulation time step (-)
$A_c$ : cross-sectional area half-way a canal (m$^2$)
$A_g$ : groundwater subsystem area (m$^2$)
$A_{gf}$ : cross-sectional area of groundwater-flow element (m$^2$)
$A_s$ : area of surface water (m$^2$)
$A_T$ : area at target level (m$^2$)
$c$ : concentration of pollutant (mg/l)
$C_c$ : Chézy coefficient (m$^{1/2}$/s)
$CM$ : control-mode set
$C_w$ : weir discharge coefficient (m$^{1/2}$/s)
$D$ : dimensionless damage (-)
$D_r$ : decay rate for pollutant (s$^{-1}$)
$E_0$ : open-water evaporation (mm)
$E_d$ : expected duration of water-system failure (h/y)
$E_f$ : expected frequency of a particular water-system failure (y$^{-1}$)
$E_r$ : reference evaporation (mm/d)
$f_A(h)$ : area function, depending on water level (m$^2$)
$f_A(V)$ : area function, depending on volume (m$^2$)
$F_F$ : flushing factor of the surface-water subsystem (d$^{-1}$)
$f_h(V)$ : water-level function, depending on volume (m+ref)
$FS$ : failure scenario
$f_V(h)$ : volume function depending on water level (m$^3$)
$g$ : gravitational constant (m/s$^2$)
$h_a$ : water level in an abstraction area (m+ref)

$h_b$      :   surface-water bottom level (m+ref)

$h_c$      :   average canal-bottom level (m+ref)

$H_c$      :   average water depth in a canal (m)

$h_{cr}$     :   weir-crest level (m+ref)

$h_d$      :   water level in a delivery area (m+ref)

$h_{dr}$     :   drain level (m+ref)

$h_{el}$     :   surface-water embankment level (m+ref)

$h_g$      :   groundwater level (m+ref)

$h_m$      :   matric pressure (cm)

$h_s$      :   surface-water level (m+ref)

$h_{sa}$     :   surface-water level in an abstraction area (m+ref)

$h_{sd}$     :   surface-water level in a delivery area (m+ref)

$h_{sl}$     :   soil-surface level (m+ref)

$H_{sl}$     :   total opening of sluice gates (m)

$h_T$      :   target water level (m+ref)

$K$       :   pollution flux (kg/s)

$K_{gf}$     :   hydraulic soil conductivity for groundwater flow element (m/s)

$k_N$      :   Nikuradse roughness coefficient (m)

$K_u$      :   uncontrolled pollution influx (kg/s)

$l$        :   linear function describing a damage-function line segment

$L$       :   total length in the surface-water subsystem (m)

$L_c$      :   canal length (m)

$L_{gf}$     :   length of groundwater flow element (m)

$n_c$      :   total side slope of canal (-)

$n_t$      :   total side slope of surface-water subsystem (-)

$p$       :   penalty coefficient

$P$       :   precipitation (mm)

$P_c$      :   wet perimeter of a canal (m)

$P_F$      :   preservation factor (-)

$P_I$      :   performance index (-)

$Q_f$      :   inflow via modeled flow elements (m$^3$/s)

$Q_g$      :   groundwater discharge (m$^3$/s)

$q_{ge}$     :   uncontrolled external flow to the groundwater subsystem (mm/h)

$Q_{gf}$     :   groundwater flow (m$^3$/s)

$Q_{gga}$    :   inflow via groundwater flow elements from other areas (m$^3$/s)

$Q_{il}$     :   flow through inlet (m$^3$/s)

$Q_p$      :   flow through pumping station (m$^3$/s)

$q_{se}$     :   external inflow into the surface-water subsystem (mm/h)

$Q_{sl}$     :   sluice discharge (m$^3$/s)

$Q_{us}$     :   surface-water inflow to prevent the water level from falling below bottom (m$^3$/s)

$Q_w$      :   discharge over a weir (m$^3$/s)

$R$       :   hydraulic radius (m)

$S$       :   sink term (d$^{-1}$)

$S_d$      :   duration of system failure (h/y)

$S_f$      :   frequency of water-system failure (y$^{-1}$)

| | | |
|---|---|---|
| $S_{rd}$ | : | drain resistance (s) |
| $S_{rg}$ | : | soil resistance (s) |
| $t$ | : | optimization time step (-) |
| $T$ | : | control horizon (-) |
| $u$ | : | control variable |
| $v_c$ | : | flow velocity in a canal (m/s) |
| $V_s$ | : | surface-water volume (m$^3$) |
| $V_T$ | : | volume at target level (m$^3$) |
| $w_c$ | : | width of canal at its bottom level (m) |
| $w_{sl}$ | : | width of one sluice sliding gate (m) |
| $w_w$ | : | width weir-crest (m) |
| $x$ | : | water-system state variable |
| $x_k$ | : | key variable |
| $z$ | : | soil depth, measured from the surface, with upward positive direction (cm) |
| $Z$ | : | objective function (-) |

# References

Assem, S. van den (1989). Bruikbaarheid van gecombineerde regenmeter-, radar- en satellietwaarnemingen. *Neerslagmeting en -voorspelling, Toepassing van moderne technieken zoals radar- en satellietwaarnemingen*. CHO Rapport Nr. 21, 43-56.

Åström, K.J. & Wittenmark, B. (1984). *Computer controlled systems; Theory and design*. Prentice-Hall Inc., Englewood Cliffs, NJ, USA.

Babovic, V. (1991). *Applied Hydroinformatics; A control and advisory system for real-time applications*. IHE Report Series 26.

Bakel, P.J.T. van (1986). *Planning, design and operation of surface water management systems; a case study*. ICW, Wageningen, NL.

Bakker, K.; Hartong, H. & Bentschap Knook, L.A. (1984). Computergesteunde besturing van rioolgemalen in West-Friesland-Oost (1). *H₂O, Tijdschrift voor watervoorziening en afvalwaterbehandeling* (17), Nr. 10, 204-207.

Barritt-Flatt, P.E. & Cormie, A.D. (1991). Implementing a Decision Support System for Operations Planning at Manitoba Hydro. *Decision Support Systems, Water Resources Planning*. Springer-Verlag Berlin Heidelberg, NATO ASI series G: Ecological Sciences, Vol. 26, 357-374.

Benoist, A.P.; Brinkman, A.G.; Diepenbeek, P.M.J.A; Waals, J.M.J. (1997). BEKWAAM, een simulatiemodel voor bekkenwaterkwaliteit. *H₂O, Tijdschrift voor watervoorziening en afvalwaterbehandeling* (30), Nr. 9, 290-295.

Beken, A. van der (1979). Neerslag meting met behulp van radar. *Ware tijd hydrologie*, 269-281. Vrije universiteit Brussel, Dienst hydrologie en het centrum voor statistiek en operationeel onderzoek.

Belmans, C.; Wesseling, J.G. & Feddes, R.A. (1983). Simulation model of the water balance of a cropped soil: SWATRE. *Journal of Hydrology*, 63, 271-286. Elsevier Science, Amsterdam, NL.

Berg, van den W.D. (1989). Enkele notities aangaande de studiedag van 16 november. *Neerslagmeting en -voorspelling, Toepassing van moderne technieken zoals radar- en satellietwaarnemingen*. CHO Rapport Nr. 21, 122-123.

Bertsekas, D.P. & Tseng, P. (1988a). Relaxation methods for minimum cost ordinary and generalized network flow problems. *Operations Research*, 36 (1), 93-114.

Bertsekas, D.P. & Tseng, P. (1988b). The relax codes for linear minimum cost network flow problems. *Annals of Operations Research*, 13, 125-190.

Botterhuis, T. (1997). *Optimalisatie van de peilhandhaving gedurende en kort na een zware bui; toepassing van neerslagvoorspelling in het operationele waterbeheer van Oostelijk en Zuidelijk Flevoland.* MSc Thesis. Delft University of Technology, Fac. Civil Engineering, Dep. Watermanagement, Environmental and Sanitary Engineering.

Boulos, P.F.; Altman, T.; Jarrige P-A. & Collevati, F. (1995). Discrete simulation approach for network-water-quality models. *Journal of Water Resources Planning and Management*, 121 (1), 49-60.

Bouwmans J.M.M. (1994). *Problematiek, normen en knelpunten bij ontwerpen waterbeheersingsplannen (discussienota waterbeheersing).* Landinrichtingsdienst.

Buat, L.G. Du (1816). Principes d' Hydraulique. Paris, France.

Buishand, T.A. & Brandsma, T. (1996). *Rainfall generator for the Rhine Catchment.* TR-183. Ministerie van Verkeer en Waterstaat, KNMI.

Cabot Plé, J. (1990). Feasibility of a RTC System for the Drainage Network of Barcelona City. *Applications of operation research to real time control of water resources systems.* EAWAG, Switzerland.

Chung, F.I.; Archer, M.C. & Vries, J.J. de (1989). Network flow algorithm applied to California Aqueduct simulation. *Journal of Water Resources Planning and Management*, 115 (2), 131-147.

Colenbrander, H.J. (1989), Editor. *Water in the Netherlands.* CHO/TNO.

Crawley P.D. & Dandy, G.C. (1993). Optimal operation of multiple-reservoir system. *Journal of Water Resources Planning and Management*, 119 (1), 1-17.

Daan, H. (1993). *Verificatie weersverwachtingen 1955 - 1993.* TR-159. Ministerie van Verkeer en Waterstaat, KNMI.

Delfland (1991). *Normen voor de bepaling van de bemalingscapaciteit, berging en peilstijging van een polder.* Technische Dienst, Hoogheemraadschap van Delfland.

Diba, A.; Louie, P.W.F.; Mahjoub, M. & Yeh, W. W-G. (1995). Planned operation of large scale water-distribution system. *Journal of Water Resources Planning and Management*, 121 (3), 260-269.

Donker, A.W. (1989). Een verificatie-experiment voor kansverwachtingen voor de neerslagsom. *Neerslagmeting en -voorspelling; toepassing van moderne technieken zoals radar- en satellietwaarnemingen.* CHO/TNO rapporten en nota's No 21.

Einfalt, T.; Grottker, M. & Schilling, W. (1990). *Applications of operation research to real time control of water resources systems.* EAWAG, Switzerland.

Esat, V. & Hall, M.J. (1994). Water resources system optimisation using genetic algorithms. *Proceedings of the Hydroinformatics'94 conference*, 225-231. Balkema, Rotterdam, NL.

Feddes, R.A. (1987). Crop factors in relation to Makkink reference-crop evapotranspiration. *ICW technical bulletins.* No 67.

Feddes, R.A.; Kabat, P.; van Bakel, P.J.T.; Bronswijk, J.J.B. & Halbertsma, J. (1988). Modelling soil water dynamics in the unsaturated zone - state of the art. *Journal of Hydrology* 100, 69-111. Elsevier Science, NL.

Franke, P-G. (1970). Abfluß über Wehre und Überfälle. *Abriss der Hydraulik*. Bauverlag Weibaden und Berlin, FRG.

Geldof, G. & Steketee, C.A. (1993). Automatisering van het waterbeheer op de Linge. *Waterschapsbelangen*, Nr. 12, 406-411.

Genuchten, M.Th. van (1980). A closed form equation for predicting the hydraulic conductivity of unsaturated soils. *Soil Science Society of America Journal*. Vol. 44, 892-898.

Goldberg, D.E. (1989). *Genetic algorithms*. Addison Wesley Publishing Company, NY, USA.

Grotentraast (1992), Chairman. *Cultuur technisch vademecum*. Vereniging voor Landinrichting.

Hafkenscheid, L.M. (1988). Numerieke neerslagverwachtingen. *Neerslagmeting en - voorspelling, Toepassing van moderne technieken zoals radar- en satellietwaarnemingen*. CHO Rapport Nr. 21, 69-73.

Hartong, H; van de Poll, C.G.; Surink, J.A. (1985). Computergesteunde besturing van rioolgemalen in West-Friesland-Oost (2). *$H_2O$, Tijdschrift voor watervoorziening en afvalwaterbehandeling* (18), Nr. 16, 114-118.

Hartong, H.J.G. (1990). Automatisering van het peilbeheer in de Flevopolders. *Sturing in het Waterbeheer*, Rapport Nr. 6, SAMWAT.

Hartong, H.J.G. & Lobbrecht, A.H. (1992). *Applications of Operations Research to Real-Time Control of Water Resources Systems*. Proceedings of the Third European Junior Scientist Course. PREDICT.

Hillier, F.S. & Lieberman, G.J. (1990). *Introduction to operations research*. McGraw Hill, NY, USA.

Hoogendoorn, G.A. (1996). A Road To Optimal Control; *Application of Linear Programming in Operational Water Management*. MSc Thesis. Delft University of Technology, Fac. Mathematics and Informatics, Dep. of Statistics, Stochastics and Operations Research.

Hooghart, J.C. (1987). *Evaporation and Weather*. CHO / TNO, Verslagen en Mededelingen Nr. 39.

Janssens, J.Ph.M.J. (1994). *Peilbeheer van de Braakmankreek; onderzoek naar de neerslag-afvoerrelatie van het gebied rond de Braakmankreek en de besturing van de Braakmanspuisluis*. MSc thesis. Delft University of Technology, Fac. Civil Engineering, Dep. Watermanagement, Environmental and Sanitary Engineering.

Jensen, M.E.; Burman, R.D. & Allen, R.G. (1990). *Evapotranspiration and Irrigation Water Requirements*. ASCE Manuals and Reports on Engineering Practice, No 70, NY, USA.

Jensen, P.A. & Barnes, J.W. (1980). *Network flow programming*. John Wiley & Sons, NY, USA.

Jilderda, R.; Meijgaard, E. van & Rooy, W. de (1995). *Neerslag in het stroomgebied van de Maas in januari 1995*, TR-178. Ministerie van Verkeer en Waterstaat, KNMI.

Khelil, A & Schneider, S. (1991). Development of a control strategy to reduce combined sewerage overflows: the case of Bremen left of the Weser. *Water Science Technology*. Vol. 24, No 24, 6, 201-208.

Klaassen, W. (1989). Neerslagwaarneming met radar. *Neerslagmeting en -voorspelling, Toepassing van moderne technieken zoals radar- en satellietwaarnemingen*. CHO Rapport Nr. 21, 11-20.

KNMI/CHO (1988). *Van Penman naar Makkink; een nieuwe berekeningswijze voor klimatologische verdampingsgetallen*. Rapport Nr. 19, CHO/TNO.

Kuczera, G. (1989). Fast multireservoir multiperiod linear programming models. *Water Resources Research*, 25 (2), 169-176.

Laat, P.J.M. de (1980). *Model for unsaturated flow above a shallow water-table; applied to a regional sub-surface flow problem*. Centre for Agricultural Publishing and Documentation, Wageningen, NL.

Lobbrecht, A.; Hartong, H. & Snabel, K. (1989). Real-time control of water management for the Flevopolder, the Netherlands. *Computational Modelling and Experimental Methods in Hydraulics*, 366-382. Elsevier Applied Science, Essex, UK.

Lobbrecht, A.H.; Hartong, H.J.G.; Landheer, O.P. & Snabel, K.W. (1990). Automatisering van de bemaling in de Flevopolders. *Waterschapsbelangen, tijdschrift voor waterschapsbestuur en waterschapsbeheer*, Nr. 15.

Lobbrecht, A.H. & Hartong, H.J.G. (1991). Real-time control of water management systems: developments in automated water level control, the Netherlands. *Advances in Water Resources Technology*, 651-661. Balkema, Rotterdam, NL.

Lobbrecht, A.H.; Verbrugge, P.P. & Hoogeveen, A.M.E. (1992). Sturing in het waterbeheer: peilvoorspelling als optimale regeling van een waterbeheersysteem. *H₂O, Tijdschrift voor watervoorziening en afvalwaterbehandeling*, Nr. 2.

Lobbrecht, A.H.; Verbrugge, P.P. (1992). Geautomatiseerd peilbeheer in oostelijk en zuidelijk Flevoland; een analyse van operationele en beheerstechnische veranderingen. *Waterschapsbelangen, tijdschrift voor waterschapsbestuur en waterschapsbeheer*, Nr. 20.

Lobbrecht, A.H. (1992). Feed-forward water-level control in rural water management systems. *Applications of Operations Research to Real Time control of Water Resources Systems*. Proceedings of the Third European Junior Scientist Course. PREDICT.

Lobbrecht, A.H. (1993a). Operational water management in rural areas. *IWSA Year Book 1993*, 309-316. International Water Supply Association.

Lobbrecht, A.H. (1993b). *Sturing in het Waterbeheer; Praktijkonderzoek Delfland; Analyse van hydrologische gegevens t.b.v. het opstellen van een model van de boezem*. Delft University of Technology, Fac. Civil Engineering, Dep. Watermanagement, Environmental and Sanitary Engineering.

Lobbrecht, A.H. (1994a). Operations modelling for control of rural water-management systems, *Advances in Water Resources Technology and Management* (ECAWART'94), 295-299. Balkema, Rotterdam, NL.

Lobbrecht, A.H. (1994b). A network flow model for optimized control of rural water systems, *Proceedings of the Hydroinformatics'94 conference*, 161-167. Balkema, Rotterdam, NL.

Lobbrecht, A.H. (1994c). *Sturing in het waterbeheer; optimalisatie in operationele waterbeheersing; Inventarisatie*. STOWA, Nr. 94-10.

Lobbrecht, A.H.; Segeren, W.A. & Nelen, A.J.M. (1995). Naar dynamische sturing in het integrale waterbeheer. *Het Waterschap*, Nr. 22.

Lobbrecht, A.H.; Segeren, A.G.; Verbrugge, P.P. & Wentholt, L.R. (1995). Technische automatisering bij regionale waterbeheerders: een enquête. *Automatisering van de waterbeheersing*. STOWA Rapport Nr. 10.

Lobbrecht, A.H. & Vredenberg, A.J. (1995). *Optimized Control of Multi-Purpose Water Systems; Determination of Operational Strategies by Integrated Simulation and Successive Linear Programming*. Working paper. Delft University of Technology, Fac. Civil Engineering, Dep. Watermanagement, Environmental and Sanitary Engineering.

Loucks, D.P.; Stedinger, J.R. & Haith, D.A. (1981). *Water Resources Systems Planning and Analysis*, Pretince-Hall Inc., NY, USA.

Loucks, D.P. (1989), Editor. *System Analysis for Water Resources Management; Closing the Gap Between Theory and Practice*. Proceedings of the Baltimore symposium (USA), IAHS, Publication 180.

Loucks, D.P. & da Costa, J.R. (1990). *Decision Support Systems; Water Resources Planning*. Springer Verlag Heidelberg-Berlin. NATO ASI series G: Ecological Sciences, Vol. 26.

Loucks, D.P (1991). Computer-Aided Decision Support in Water Resources Planning and Management. *Decision Support Systems, Water Resources Planning*. NATO ASI series G: Ecological Sciences, Vol. 26, 3-41. Springer Verlag Heidelberg-Berlin.

Luijendijk, J. & Sinke, M.D. (1982). Hoogheemraadschap van Delfland: Veranderingsprocessen in de waterhuishouding. *PT Civiele Techniek*, 37, Nr. 9.

Lumadjeng, H.S. (1985). *Een eerste aanzet tot een ontwerp van een lineair optimaal regelsysteem ten behoeve van het stuwbeheer*. Prov. Gelderland, Dienst WB, afdeling waterhuishoudkundig onderzoek.

Makkink, G.F. (1957). Testing the Penman formula by means of lysimeters. *Journ. Int. of Water Eng.*, 11, 277-288.

Male, J.W. & Soliman, K. (1981). Lake Michigan diversion for water quality control. *Journal of the Water Resources Planning and Management Division*, ASCE, 107 (WR1), 121-137.

Mariño, M.A. & Mohammadi, B. (1983). Reservoir operation by linear and dynamic programming. *Journal of Water Resources Planning and Management*, 109 (4), 303-319.

Martin, Q.W. (1987). Optimal daily operation of surface-water systems. *Journal of Water Resources Planning and Management*, 113 (4), 453-470.

Martín-García, H.J. (1996). *Combined logical-numerical enhancement of real-time control of urban drainage networks*. Balkema, Rotterdam, NL.

Mays, L.W. (1989). Hydrosystems engineering simulation vs. Optimization: why not both?. *Proceedings of the Baltimore Symposium*, 225-232.

Mészáros, Cs. (1996). *The efficient implementation of interior point methods for linear programming and their applications*. PhD thesis. Eotvos Lorand University of Sciences, Hungary.

Molen, W.H. van der (1992), Editor. *Richtlijnen voor het berekenen van afwateringsstelsels in landelijke gebieden*. Werkgroep afvoerberekeningen, Landinrichtingsdienst.

Nelen, A.J.M. (1992). *Optimized control of urban drainage systems*. PhD thesis. Delft University of Technology, Fac. Civil Engineering, Dep. Watermanagement, Environmental and Sanitary Engineering.

Neugebauer, K.; Schilling, W. & Weiss, J. (1991). A network algorithm for the optimum operation of urban drainage systems. *Water Science Technology*, 24 (6), 209-216.

NWRW (1989). *De vuiluitworp van gemengde rioolstelsels*. STORA / Ministerie van VROM.

Penman, H.L. (1948). Natural evaporation from open water, bare soil and grass. *Royal Society of London, Series A*, 193, 120-145.

Pezeshk, S.; Helweg, O.J. & Oliver, K.E. (1994). Optimal operation of groundwater supply distribution systems. *Journal of Water Resources Planning and Management*, 120 (5), 573-586.

Pike R.W. (1986). *Optimization for engineering systems*. Van Nostrad Reinhold, NY, USA.

Roos, C. (1987). *Netwerkmodellen en algoritmen; lecture notes a86H*. Delft University of Technology, Fac. Mathematics and Informatics, Dep. of Statistics, Stochastics and Operations Research.

Rooy, W.C. de & Engeldal C.A. (1992). *Neerslagverificatie LAM*, TR-143. Ministerie van Verkeer en Waterstaat, KNMI.

Salland (1995). *Integraal watervoorzieningsplan Luttenberg*, Waterschap Salland.

Schilling, W. & Petersen, S.O. (1987). Real time operation of urban drainage systems - validity and sensitivity of optimization techniques. *Systems Analysis in Water Quality Management*. Pergamino, Oxford, UK.

Schilling, W. (1989), Editor. *Real time control of urban drainage systems, The state-of-the-art*. IAWPRC Scientific and Technical Reports No 2. Pergamino, Oxford, UK.

Schilling, W. (1991). Real-time Control of Urban Drainage Systems: From Suspicious Attention to Wide-Spread Application. *Advances in Water Resources Technology*, 561-576. Balkema, Rotterdam, NL.

Schilling, W. (1996). Real Time Control of Waste Water Systems (Malmö-Klagshamm RTC system). *Journal of Hydraulic Research*, Vol. 34, No 6, 785-797.

Schuilenburg, D.J. (1996). *Verdrogingsbestrijding in Luttenberg, Waterschap Salland; Schadebeperking aan Natuur en Landbouw door een verbeterd Beheer*. MSc thesis. Delft University of Technology, Fac. Civil Engineering, Dep. Watermanagement, Environmental and Sanitary Engineering.

Schultz, E. (1992). *Waterbeheersing van de Nederlandse droogmakerijen*. Van Zee tot Land 58. Directoraat-Generaal Rijkswaterstaat.

Schultz, G.A. (1989). Ivory tower versus ghosts? - or - the interdependency between systems analysts and real-world decision makers in watermanagement. *Systems Analysis for Water Resources Management; Closing the Gap Between Theory and Practice*. IAHS Publication No 180.

Schuurmans, J. (1997). *Control of Water Levels in Open Channels*. Draft PhD thesis. Delft University of Technology, Fac. Civil Engineering, Dep. Watermanagement, Environmental and Sanitary Engineering.

Segeren, W.A. (1983). Introduction to Polders of the World. Keynote. *Water International*, 8, 51-54, Elsevier Sequoia, NL.

Segeren, W.A. (1966). Drainage requirements of newly reclaimed marine clay sediments as influenced by subsoil conditions, Question 21, R6. *Transactions of the 6th congress, International Commission on Irrigation and Drainage (ICID)*, New Delhi, India.

Smakman, J.C. (1993). *Optimization algorithms for complex river basin systems*. MSc thesis. Delft University of Technology, Fac. Mathematics and Informatics, Dep. of Statistics, Stochastics and Operations Research.

Smedema, L.K. & Rycroft, D.W. (1983). *Land drainage; planning and design of agricultural drainage systems*. Batsford Academic and Educational Ltd. London, UK.

Spaan (1994). *De Q-h relatie van de Crump-De Gruyter onderspuier*. Delft University of Technology, Fac. Civil Engineering, Dep. Watermanagement, Environmental and Sanitary Engineering.

Sprague, R.H. & Carlson, E.D. (1982). *Building Effective Decision Support Systems*. Englewood Cliffs, Prentice-Hall, New Jersey, USA.

Steenbekkers, J.A.H. (1996). *Het waterbeheer van Delflands boezem; Infrastructurele veranderingen in het beheersgebied van het Hoogheemraadschap van Delfland en de consequenties voor het operationele waterbeheer*. MSc thesis. Delft University of Technology, Fac. Civil Engineering, Dep. Watermanagement, Environmental and Sanitary Engineering.

Straten, G. van; Bots, W.C.P.M. & Riel, P.H. van (1986). Waterkwaliteitsbeheersing door afvoersturing in het watergangenstelsel rondom Almelo. *H₂O, Tijdschrift voor watervoorziening en afvalwaterbehandeling*, Nr. 1.

Sun, Y-H; Yeh, W.W-G; Hsu, N-S & Louie, P.W.F. (1995). Generalized Network Algorithm for Water-Supply-System Optimization. *Journal of Water Resources Planning and Management*, 121 (5), 392-398.

Terlaky, T. (1996), Editor. *Interior point methods of mathematical programming*. Kluwer Academic, Dordrecht, NL.

Thomann, R.V. & Mueller, J.A. (1987). *Principles of Surface Water Quality Modeling and Control*. Harper Collins, NY, USA.

Turban, E. (1990). *Decision Support and Expert Systems; management support systems*. Macmillan International, NY, USA.

Ven, F.H.M. van de (1983). Duurlijnen; gebruik en misbruik. *Cultuurtechnisch Tijdschrift*, (23), 1-8.

Ven, F.H.M. van de (1989). *Van neerslag tot rioolinloop in vlak gebied*. Van Zee tot Land Nr. 57. Ministerie van Verkeer en Waterstaat, Rijkswaterstaat, directie Flevoland, Lelystad.

Ven, G.P. van de (1993). *Man-made lowlands, History of water management and land reclamation in the Netherlands*. Matrijs, NL.

Verbrugge, P.P. & Lobbrecht, A.H. (1993). Automated water-level control in the Flevopolders (NL): Analysis of effects on operational water management. *Proceedings of the 15th congress of the International Commission on Irrigation and Drainage (ICID)*, Question 44, Volume 1-A, R.23, 271-282.

Vermeulen, C-J.M. (1992). Anticipatory control for a spilling sluice. *Journal A*: Vol. 33, No 3, 94-98.

Vermeulen, C-J.M.; Hartong, H.J.G.; van Kruiningen, F.E. & Moser, G.M. (1994). Real-time water management of the Rijnland storage basin. *Advances in Water Resources Technology and Management*, 327-333. Balkema, Rotterdam, NL.

Verspuy, C. & Vries, M. de (1981). *Lange Golven, Lecture Notes b73*. Delft University of Technology, Fac. Civil Engineering, Dep. Fluid Mechanics.

Versteeg, R.P. (1997). *Verdrogingsbestrijding in het stroomgebied van de Aa of Weerijs*. MSc Thesis. Delft University of Technology, Fac. Civil Engineering, Dep. Watermanagement, Environmental and Sanitary Engineering.

Vredenberg, A.J. (1996). *Mathematical optimization methods in Dutch operational water control; Applicability of common methods*. MSc thesis. Delft University of Technology, Fac. Mathematics and Informatics, Dep. of Statistics, Stochastics and Operations Research.

Wartema L. (1989). Neerslagmeting en neerslagvoorspelling; wensen van de Nederlandse landbouw. *Neerslagmeting en -voorspelling, Toepassing van moderne technieken zoals radar- en satellietwaarnemingen*. CHO Rapport Nr. 21, 118-119.

Werkgroep Hydrologisch Onderzoek Overijssel (1976). *Hydrologisch onderzoek Salland.*

Wösten, J.H.M.; Veerman, G.J. & Stolte, J. (1994). *Waterretentie- en doorlatendheidskarak- teristieken van boven- en ondergronden in Nederland; de Staringreeks.* DLO Staring Centrum, TD 18.

Wright, J.L. (1982). New evapotranspiration crop coefficients. *Journal of Irrigation and Drainage.* ASCE, 108 (IR2), 57-74.

Wurbs, R.A. (1993). Reservoir-system simulation and optimization models. *Journal of Water Resources Planning and Management,* 119 (4), 455-472.

Yeh, W.W-G. (1985). Reservoir management and operations models; a state-of-the-art review. *Water Resources Research,* 21 (12), 1797-1818.

Zeeuw, J.W. de & Hellinga, F. (1958). Neerslag en afvoer. *Landbouwkundig tijdschrift.* No 70, 405-422.